23 July 2026

DINNER WITH AN ASTRONAUT

DINNER WITH AN ASTRONAUT

SERVING SPACE STORIES: PAST, PRESENT AND FUTURE

LEROY CHIAO

WITH VICTORIA BRUCE

HANOVER SQUARE PRESS

HANOVER SQUARE PRESS™

Recycling programs for this product may not exist in your area.

ISBN-13: 978-1-335-00067-5

Dinner with an Astronaut

Hanover Square Press
22 Adelaide St. West, 41st Floor
Toronto, Ontario M5H 4E3, Canada
HanoverSqPress.com

HarperCollins Publishers
Macken House, 39/40 Mayor Street Upper,
Dublin 1, D01 C9W8, Ireland
www.HarperCollins.com

Printed in U.S.A.

26 27 28 29 30 LBC 5 4 3 2 1

To my parents, T.T. and Cherry Chiao.
Thank you for shaping me into who I became.

DINNER WITH AN ASTRONAUT

CONTENTS

FOREWORD

DINNER PLANS

LIKE FLYING INTO SPACE, sitting down and sharing a meal with friends and loved ones has always been one of my favorite ways to spend time. When I look back at my life, I see it like a long, winding conversation over dinner. Maybe because of my Chinese DNA and appreciation for everything that came before me, I don't see my astronaut career in a vacuum. I never considered my travel to space a solo accomplishment. I'm not Leroy Chiao, superhero, donning a spacesuit with a patch of the American flag. Instead, the story of how I got to space runs the course of history like a well-seasoned meal, my friends and family and colleagues sat at the dinner table with me, all of us telling tales of ingenuity, brilliance, intrigue, and wonder.

Was there ever a time that humans didn't look up and marvel at the dark sky and the bright Moon circling around the backdrop of stars? I don't think so. I'm guessing Neanderthal families, dining by campfires, tried to find meaning up there. Humans are curious. At dinner as a kid, my family talked about engineering, science, and the war that drove my parents from their homes. With my kids, we talk about history, technological advances, and their futures.

Nowadays, my dinner companions often ask how a nerdy kid from the Midwest became the world's first full-Chinese heritage professional astronaut. They're fascinated by the stories of the space race, and curious where space exploration goes from here. They ask me tons of question about my time suiting up and launching into space. Some people even ask: *Have you ever seen a UFO?* (Stay tuned for the answer—you may be surprised.)

Now I invite you to join me for a dinner conversation, to ask me the questions that will reveal the giants of physics, chemistry, and engineering who carried me into space, and the wild things that happened once I got there.

Bon Appetit. Chènrè chī. Dig in!

PART ONE

HORS D'OEUVRES

Whet Your Appetite with a Taste of the
Science and History of Space

Q1: WHEN DID YOU KNOW YOU WANTED TO BE AN ASTRONAUT?

IN THE SPRING OF 1969, millions of American kids like me were glued to news of the upcoming Moon launch. That's when I decided I *had* to go to space. I lived with my mom, dad, and two sisters in a new housing development in northern California. We were the only Asian family in a sea of white. To me, it was all I knew. To my parents, it was a foreign world that suited them just fine.

My friends called me Leroy, my American name. My parents called me *Érzi*, Son, or my Chinese name, *Xiǎo Zhong. Xiǎo* means little. When you're little and Chinese, *xiǎo* is basically the prefix to your name. Mom and Dad made us speak Chinese at home, and Dad fined us a nickel if he caught us speaking English.

My father came from a small farming community in Northern China, in the western part of Shandong province.

My dad always said he's against tradition, but there were a few things that tied him to the old country. 1. Education is King. 2. Never spend money when you don't have to. 3. The individual's needs are not important—especially children's. 4. Chinese food is the best.

Mom and Dad were both engineers. They believed in planning and systems. If I was hoping for a sibling close to my age to play with, it was not going to happen. Dad landed a job after grad school with the army at Rock Island Arsenal in Milwaukee, Wisconsin, and they timed us kids to come out five years apart. When my older sister turned five, I arrived. When I was a few months past four, they went to work on *Xiǎo Hai* 3.0.

We'd moved to Wichita, Kansas, for dad's job with Koch Industries just before the time Mom's belly was full with Sibling Three. Mom was at the kitchen table one morning writing a letter to China. I could tell it was going to another country because it was on flimsy airmail paper that folded into an envelope.

"*Māmā*, who are you writing to?" I asked.

"Your father's family, *Xiǎo Zhong*, to see what we should name the new baby."

Asking Chinese elders what to name your American baby sounded very traditional, which surprised me since my dad hated old customs.

"Is that how I got my name, too?" I asked. My full first name is *Li Zhong.*

"Yes, *Érzi.* Your name means 'something that lies in the middle,' 'something balanced.'"

Did the elders know I would be the middle kid? Did they know my parents' baby-making methodology? Did I feel "balanced"? I wasn't sure. I didn't ask Mom about my American name, but I figured Leroy was the closest she could find to *Li Zhong.*

My parents met in college in Taiwan after World War II, where they were both working on chemical engineering degrees. They got married and had my older sister, *Ming zi.* After graduation, Dad got into graduate school at the University of Rhode Island in the chemical engineering program. Mom came a year later and left my sister with Grandma and Grandpa in Taiwan. Mom wanted to continue with grad school and applied to the same program as Dad, but the chairman of the chemical engineering department wouldn't have it. It was 1956. There were hardly any women in the field. The chairman wanted to keep it that way. "I'll be damned if I'm going to have a woman in my engineering program," he told my dad. Mom got a master's degree in the chemistry department instead. She told me she never felt discriminated against professionally for being Chinese in America, but she did—many times—for being a woman. I didn't understand why. Mom was the smartest person I knew.

Mom never discussed feeling different than the people we lived around. But for me, I couldn't help but notice some kid-obvious contrasts growing up. In Danville, I had two best friends on my street, brothers Jack and Pete. We never discussed it. It was clear we weren't the same. My parents made them take their sneakers off to come into the house. When I went to their house, I left my shoes on, even on the carpet. My family ate with chopsticks, they used forks. Their mom stayed home, my mom went to work.

Then in 1969, something big happened that eclipsed all our differences. With millions of other American kids, Jack, Pete, and I became obsessed with rockets, spaceships, and the Moon. The hobby stores sold out of plastic models of the

Lunar Module, NASA rockets, and Apollo space capsules. The local toy store carried action-figure astronauts dressed in spacesuits with rectangular backpacks and mirrored face-masks. My friends and I cheered on Neil Armstrong, Buzz Aldrin, and Michael Collins. We booed the communist cosmonauts who beat us to orbit, and we swore we'd beat them to the Moon. I never saw any astronauts who looked like me, but I didn't care. I was obsessed.

Over the course of a few weeks, I built a spaceship out of cardboard in the garage under Dad's workbench and invited Jack and Pete over to train for a moonwalk. We played in the spaceship for hours, setting up the cockpit under my dad's workbench and converting some leftover pegboard into a control panel.

After we were done and Jack and Pete went home, I got the milk from the fridge and smelled it to make sure it wasn't bad. I silently cursed my uncle who introduced powdered milk to my parents so they could save money. I stirred in some Nestlé's Quik to hide the nastiness. *Maybe powdered milk is part of my astronaut training*, I told myself to make it easier to swallow.

At the table, Mom and Dad chatted about Mom's job at a small engineering firm and Dad's work at the Lawrence Radiation Laboratory. Dad did research on composite materials, for use in aerospace and other applications. The engineers at Lawrence combined polymer resins and fibers, then heated and pressurized them to make superstrong and lightweight structures. My parents could tell I had an affinity for engineering as I asked questions about Nomex, resin, and carbon fiber. I wanted to know how they used the stuff Dad and Mom made in the real aerospace world.

"*Xiao Zhong*, you can do these things when you get a job after college," Dad said.

I waited for a second, chewed, and swallowed my fried rice.

"Actually," I announced. "I am going to be an astronaut," I said, hoping they'd be impressed they were having dinner with a future spaceman.

Dad was silent (which was better than the alternative).

"Well, that's a bit dangerous," Mom said.

I waited for her to finish, my heart pounding.

"You have to be an engineer first, though."

I grinned big. For me, that was as good a go-ahead as it got.

Q: WHAT WAS IT LIKE TO BE AN AMERICAN KID GROWING UP WITH VERY CHINESE PARENTS?

IN 1967, WHEN I was seven, Dad got a job working for the Lawrence Radiation Laboratory in Livermore, California. We moved from Wichita to Danville, California, about thirty minutes away from Dad's work. Like our neighborhood before, pretty much all the other folks were white. It took me a while to understand I was an anomaly in the world around me. Dad came from a province in China that eats a lot of steamed bread. My parents made *bao* at home, and Dad packed my lunch with *bao-zhi,* folded puffy bread with a mixture of pork and vegetables inside. At home, I loved these. *Bao-zhi* at school was a different story.

On the playground, I jealously eyed other kids savoring peanut butter and jelly sandwiches on Wonder Bread. I listened longingly to the *crunch, crunch* of Ruffles potato chips.

The one that killed me the most, the *shreep* sound of the metal pull-top from a Heinz pudding cup.

I took out my puffy *bao-zhi*.

"Ew, Leroy. What is that weird-looking thing?" Bill asked.

"That's disgusting!" John added.

John turned his back on me, avoiding the stench of my food.

"He's such a weirdo," I heard him mumble.

I'd had it. I went home, determined to get through to my dad. I needed a regular American brown bag lunch.

"Baba, can I just have *sandwiches*? Everyone thinks our food is weird."

He may have felt my desperation to fit in, because he took pity on me.

"Okay, *Érzi*."

Dad went to the store and bought the cheapest brown bread they had. It was the consistency of Wonder Bread, but he assumed it was healthier because it was brown. I was thrilled. I went to bed excited to join the mainstream at lunch the next day.

I sat down in the cafeteria, confident that I was just like every American kid. I dug out my sandwich, triple wrapped in Saran Wrap and gasped. Sweat broke out on my forehead. The shape was all wrong. I peeled back one plastic layer, then two, then three. My dad had taken one slice of bread and wrapped it just like a *bao*—around an entire chicken leg, bone and all. No mustard, no mayo. Nothing.

My parents were from the old country. They never ate meat that wasn't overcooked because there were so many parasites; anything not well-done would make people sick. Dad bought fully cooked ham then baked it in the oven until it was dry like cardboard. We didn't have a meat slicer, so he stuck a half-inch ham brick between brown bread slices. I couldn't even bite through it. My parents bought the cheapest cuts of meat, especially if it was on sale because it was about to expire. They cooked it until it had the consistency of shoe leather. My dad occasionally bought frozen pizza, something my friends raved about. But my dad only bought off-brand items (they were cheaper) and refused to preheat the oven because, "That is just wasting energy. Wasting energy is wasting money." So, the pizzas were always soggy and gross.

All of this made me to want to have school lunches as often as my parents allowed. They usually didn't object too much, since the heavily subsidized lunches cost next to nothing. I carried my Tater Tots and mystery meat to the lunch table with pride.

My gastronomic life changed completely when Mom went back to work. My parents hired a Black woman named Nicey as a housekeeper and to watch my little sister and me after we came home from school. We hadn't seen many Black people except a few on TV since our neighborhoods in Milwaukee and Wichita were all white too. Nicey was a little older than my parents, and as nice as any adult I'd ever met. My parents were not prejudiced at all and went out of their way to make sure we grew up the same.

"You shouldn't treat anyone differently," Mom always said. "Everyone deserves respect."

Dad said that China was a very racist country. "That is not the right way to be," he told us.

That first day with Nicey was a revelation to my taste buds.

Wow, this is what a burger is supposed to taste like? And BBQ spareribs! Wow, my new favorite!

The juicy meat, pickles, tomatoes, and ketchup danced on my tongue. The soft consistency was a pleasure to chew.

The ribs were falling off the bone with such wonderful flavor. My school sandwiches now had normal lunch meat *and mustard*! It was glorious.

Nicey was not just a good cook; she was a loving caregiver. Once I had the flu and stayed home from school. I was throwing up and burning hot. I didn't even want to watch TV. The next afternoon, I started feeling better.

"Nicey," I asked, "can I have a little sandwich?"

"Of course, Leroy."

Nicey knew my tummy was in recovery, so she made me a tiny quarter sandwich.

I gobbled it up. I started to feel better and asked for another one.

"Sweetie, do you want me to make you a whole sandwich or half a sandwich?" she said.

"No, I like the little quarter sandwiches," I said. They were so cute, and bite-sized.

She happily made me little quarter sandwiches one by one until I was full. I could tell she loved me. She showed me with food.

After two years taking care of our family, Nicey announced she was going off and getting married. We were happy for her, but I was sad.

Now that my little sister was in kindergarten, and I was nine, Mom and Dad decided we didn't need another housekeeper. I was old enough to get my younger sister, Sandy, on the bus to school, back home, and make her an after-school snack. They hid a key in the garage.

I tried to fill the food vacuum that Nicey left in my world. I made spaghetti with Campbell's soup for sauce. I stirred in some American cheese slices and added tuna from a can. I created a signature noodle dish with tuna, canned mushrooms, and chicken bouillon. I confidently tossed in a pinch of garlic powder and black pepper.

Years later, after college, I learned to make my dad's famous chow mein, mom's stir-fry, and the delicious pork filling for the *jiǎo zi*. When it was my turn to cook for fellow graduate students and later for some of my astronaut colleagues, I earned a bunch of fans for making some of these dishes. Instead of being embarrassed, I was proud to share my authentic Chinese cooking with friends.

Q2: WHAT IS THE HISTORY OF CHINESE SPACE EXPLORATION?

I'M NOT SURE IF my Chinese ancestors ever envisioned humans flying spaceships or floating above Earth in bulky, segmented spacesuits. They were, however, very much in tune with the rhythms of the universe. The ancient Chinese believed our lives intertwined with celestial bodies and cycles, and that human lives should follow the cosmos. So, it's not surprising that Chinese mythology surrounding the formation of the universe is based on a human-ish looking god named Pangu.

Legend says that Pangu gestated inside a giant egg for 18,000 years. Once the contractile force of Yin and the expanding force of Yang got into perfect balance, Pangu emerged—a hairy, horned giant on a mission. He split Yin and Yang with his ax and turned the former into the Earth and the latter into the sky. Over another 18,000 years, the

Earth and sky grew and grew, crushing Pangu between the two. His bones turned into rocks and mountains, his blood into the Earth's rivers and oceans, and his hair into the stars.

My parents' young life was very different from mine. If they knew the whimsical tale of Pangu, they didn't pass it on to my sisters and me. They were children when the Japanese invaded China on the eve of the Second World War. When the enemy made it to my father's small village, everyone who could, ran. Someone lifted my father on top of a packed train where he and his family held on to each other along with a throng of neighbors. If he fell from the train speeding through the night, everyone he knew would be lost to him forever.

My mom's family had a different journey. Her father had studied electrochemistry in Japan back when relations between the two countries were better. He was a village leader who spoke fluent Japanese and understood Japanese culture. When the Japanese army came through their village, for a while, my grandfather was able to be an effective go-between and kept his village from suffering most of the atrocities the Japanese soldiers committed throughout occupied China. But soon my mom and her family had to flee too.

Both Mom and Dad spent their young lives running, first from the Japanese, then from the purges of Mao Zedong's communist regime. They escaped one place and evacuated another. They didn't have time for celebrations or traditions. There was no time for religious beliefs either, so there was no god to pray to for help. I was about the same age as my dad was during WWII when I rode my bike down our quiet cul-de-sac and knocked on Jack's door. It was Sunday. I was surprised to see my friend all dressed up.

"Where are you going?" I asked.

"We're going to church."

"Huh? How often do you go to church?"

"Every Sunday."

"For how long?"

"Well, most of the morning."

Dang, you spend an entire morning there every Sunday? I was in disbelief.

"What do you do at church?"

"We listen to the preacher."

I calculated all the model-building, bike-riding, and spaceman-playing that could be done in that prime weekend real estate. *What a bummer.*

One gloomy day, when I ran out of models to build, I went to meet the new neighbors next door. The kids were around my age, and we became friends. One day, they started talking about God this, Jesus that. I was mystified. They told me they were Mormons. I had no idea what that meant.

I went back home and found my parents cooking lunch.

"Is there a God?" I asked.

"*What?*" they said in unison, both heads snapping in my direction.

"The kids next door told me all about *God*. Is there a God?"

"No," Dad said, flatly.

There was an awkward pause, then Mom said, "You can't tell him that! We don't know, *Érzi*. We're not sure. Different people believe different things."

And that was the end of that.

The one holiday we celebrated was Chinese New Year. Unlike New Year's Day, it comes on a different day every year.

"Why is the date always different?" I asked my mom.

"It's just different," she said. "It goes by the cycle of the Moon."

It turns out that Chinese astronomers created a calendar

based on the Moon instead of the sun—what we use in the Western world. In the lunar calendar, there are twelve months that have either twenty-nine or thirty days. It makes for a year that's eleven days short of the Earth's revolution around the sun. To keep the seasons in the same month over time, the lunar calendar adds an extra month seven times every nineteen years. Mom had a way of counting on her fingers to figure out when the Lunar New Year would start.

"Can you teach me how to do it?" I asked her.

She brushed me off. "Oh, don't worry about it. It's not important." That was the old country.

Chinese people don't count time from before and after the birth of Jesus Christ (obviously), but rather from the coronation of the mythical first Chinese emperor Qin Shi Huang (the Yellow Emperor), which happened 2697 years before Christ. So basically, I was born in 1960 plus 2697 which makes my birth year 4657 in the Chinese calendar.

On Chinese New Year, we got together with my cousins on Mom's side for a bare-bones celebration in California. We didn't have fancy dragon decorations or red paper lanterns strung across the garden. Mom and Dad cooked stir-fry and the kids pinched dim sum wrappers into dumplings. When we first moved to California, Dad took us to Chinatown in San Francisco. There was no Chinatown in Wichita or Milwaukee back then, and we kids had never seen anything like it. One of my fondest memories of Chinatown was the triangle-shaped sticky-rice cakes with a sweet wine flavor. *Delicious.* Most Chinese celebrations are centered around food, which makes sense. It seems so obvious nowadays that we divide a full orbit of the Earth around the sun into twelve months and four seasons. But imagine having no calendar to mark the passage of time or predict the coming climate shifts.

Before the lunar calendar, Chinese farmers tracked stars to figure out when to plant crops. We needed rice on the table and feed for our cows, chickens, and pigs. The farmers took their cues from *Ta-Chin*, the great rulers of the sky—what the West calls Scorpio, Orion, and Ursa Major.

By the Shang dynasty (1600–around 1050 BCE), Chinese astronomers bypassed the complicated constellations and looked to the Moon to predict the growing season. When the Moon lined up with the Earth and Sun, it marked *"Shuo Ri"*, the first day of the month. The full Moon marked the month's middle.

It wasn't completely accurate, and the seasons shifted over time. Emperors needed credibility when it came to predicting weather shifts and announcing the planting season. A ruler looked especially good if he could call a coming eclipse and get it right. So kings packed their courts with astronomers who tried to keep the calendar straight throughout the year and give the emperor a heads-up on other celestial phenomena.

By the 10th century and the Song dynasty (960–1279), a brilliant scientist named Shen Kuo got the job to head China's Imperial Observatory. Shen had the goal of nailing down the calendar once and for all, as well as predicting the solstice with 100 percent accuracy, the solar and lunar eclipses, and other assorted celestial goings-on.

Shen was relentless at striving for accuracy and tireless at collecting data. In 1088, Shen magnetized iron needles and floated them in water to create a magnetic compass. After more than three years of collecting measurements and more than two hundred sketches, his beloved geometry revealed that the compass pointed to the magnetic north, not the true north pole. It was a brilliant day for navigation, exploration, and bragging rights for the Song Empire. *Fun fact:* It took the Western world another four hundred years to get up to speed on magnetism and true north.

Shen got so deep into space that the emperor asked him to lead the Imperial Observatory, where he planned to tackle some practical science. "The current calendar has fallen," he wrote. On the month-to-month, things were consistently out of whack, and it appeared the team of imperial astrologers had some dodgy systems to try to right the wrongs. "For example, the algorithm uses red chips and black chips to distinguish between positive and negative numbers," Shen wrote. "The calendar does not understand the meaning of this naming and thinks that there are nine tracks in the operation of the Moon. This is ridiculous."

The calendar was causing embarrassing political problems too. The country was enormous, and some calendar updates never made it to far away territories. When foreign leaders made pilgrimages to the Imperial City to celebrate the solstices, they often showed up at the wrong time. *Awkward.*

Shen proposed his fix. They would collect accurate astronomical observations of the Moon and planets by making three nightly checks over the course of five years. This was unfortunate news to the astronomers at the observatory. They were political appointees and had no desire to lose sleep doing boring data collection.

Shen had a hard time as a boss. Since nobody at the observatory liked him or would take his direction, he brought in his own guy, Wei Pu, a commoner with a brilliant mind for astronomy, chronometry (the study of time), and new ideas, which were hard to come by around the observatory. Wei realized that the old technique for finding the position of the sun wasn't accurate. He cleared that up with complicated math (Shen bragged that Wei did these calculations in his head) and added the sun's position to eclipse theory, upping the accuracy of predictions and angering the old guard.

Wei and Shen presented their data in writing and even made a public display, proving their hypothesis on lunar motion. The ministers in charge of the calendar made a couple of changes with the new data but ignored most of the new calculations. Shen was livid. "How sad that the backbiting of that bunch of calendar-makers could have kept [Wei] from bringing his art to fruition!" The wheels of human inertia and envy grounded Shen and Wei's project to a halt, and the old calendar remained in place to puzzle men for centuries to come.

Q3: WHAT ABOUT AMERICANS WHO STUDIED SPACE FROM THE EARTH?

AS AN AMERICAN HIGH SCHOOL kid in California, I didn't learn about Shen Kuo or any contributions to science by Chinese researchers. My high school textbooks were filled with Greeks, Italians, Germans, Brits, and Americans.

But I was still super interested in science and technology from early on. I asked tons of questions about everything and read every science book I got my hands on. In Kansas, where we moved when I was three, we lived in a rural area where we had a sky full of stars.

One fall evening when I was about five, we were coming back from dinner. Mom took my hand as I jumped out of the back seat. The sky was especially clear.

"Look up there, *Érzi*," she said, waving her hand from the

horizon upward. "All of this is called the Milky Way. This is our galaxy."

I looked up, scanning the vastness.

"Each one of those is a star," she continued. "Each star is like our sun. A bunch of stars together make up a galaxy."

"Ah," I said. "They're just far away. That's why they look so small. Right?"

I didn't know it at the time, but just before the turn of the century, not far from my Kansas home, another midwestern boy had been looking up in awe at the Milky Way. While I would go on to love planes, rockets, and ships that would take humans into orbit, Edwin Hubble only ever got to travel to space through the lens of a telescope.

As a boy in the 1890s, Edwin received a telescope built by his grandfather—a present for his eighth birthday. Later, he spent thousands of hours looking through the world's largest telescope in Southern California. What he found would be the greatest astronomical discovery since Galileo proved Copernicus right, that the Earth rotates around the sun.

Like Shen Kuo and me, Hubble was addicted to understanding why things are the way they are. "Equipped with five of his senses," Hubble famously wrote, "man explores the universe around him and calls the adventure Science." The six-foot-three, movie-star-handsome Hubble was not a typical nerdy scientist. In high school, he dominated in football, basketball, track, and boxing. With endless charm and a brilliant mind, he sailed through school and was off to college at sixteen. He may have dreamed of a career in astronomy, but his businessman father decided that he needed to go to law school.

He got a Rhodes scholarship, went to Oxford, and painfully studied law, which bored him to death. The stars were calling him, so Hubble went back for a PhD in astronomy at the University of Chicago. There, at the Yerkes Observatory on campus, he tracked nebulae—bright cloudy areas of gas and dust—and created a system to classify the different types. His ideas gained traction, and his charisma earned him a lot of fans. On the eve of finishing his PhD, he got a job offer from George Ellery Hale, the famous astrophysicist running the Mount Wilson Observatory, in Pasadena, California. It was a dream job.

But it was 1917, and there were more pressing things here on Earth. Hubble sat up all night, rushed to finish his thesis, got a waiver to bump up his oral exam to the next morning, then packed his bag. The Hooker Telescope, the most powerful in the world at Mount Wilson, would have to wait. Hubble telegraphed Hale: "Regret cannot accept your invitation. Am off to the war."

Sadly for Hubble, who scored a command of an infantry with several stellar letters of recommendation from his professors, he saw no action before the end of the war. He traveled around Europe for a while, making connections and checking out European observatories. In 1919, Hubble landed in San Francisco. Still in his officer uniform, he traveled to Mount Wilson in Pasadena to meet the giant metal and glass creature he would become intimately linked to for the rest of his life.

There were two astronomical theories that gnawed at Hubble. Standard school of thought was that the Milky Way encompassed the entire universe. Everything out there was part of the same galaxy—*our* galaxy. The other theory was that the universe was static and unchanging. This one was rock solid in the dogma. In 1916, a thirty-seven-year-old theoretical physicist named Albert Einstein calculated that the universe was expanding. He got so much clapback from the establishment, he added a "cosmological constant" to his theory of relativity, changed his tune, and shuffled back to the static universe camp.

About four years into his work at Mount Wilson, Hubble noticed a nova star standing out brilliantly in a "nebula" called Andromeda. The variable star was a Cepheid, a star that changes in intensity over time. With the Cepheid, Hubble could measure distance. He found that the Andromeda "nebula" was not in our galaxy at all. In fact, it was more than a million light-years away. And it was not a nebula. It was its very own galaxy with billions of stars of its own.

In a very unscientific but highly American move, Hubble hired a publicist to lobby for a Nobel Prize. A slew of media hits with photos of handsome Hubble at the Hooker and catchy headlines like "Behold the Universe!" and "Pilgrimage to Eternity" hit the newsstands. *Travel* magazine gushed, "While Columbus sailed three thousand miles and discovered one continent and some islands, Hubble has roved through infinite space and discovered hundreds of vast new worlds, islands, subcontinents, and constellations not just a few thousand miles away, but trillions of miles out yonder." Unfortunately for Hubble, there was no Nobel Prize for astronomy, and the selection committee was unmoved by Hubble's newfound celebrity to change their policy.

Hubble never tried to link his discovery to the origin of the universe or the formation of the galaxies. Big bang theory, which couldn't have come without him, came later by other researchers. So, unlike Galileo, Edwin Hubble didn't remain on the tip of American high school science teachers' tongues into the future. Even as a kid obsessed with space, I wouldn't hear his name until decades later, when NASA built a giant telescope and named it after him.

Hubble died in 1953 of an aneurysm that ceased all the discovery in his brilliant brain. Five years later, President Dwight Eisenhower launched the United States' first astronaut program, Project Mercury. Americans were on the verge of getting up close and personal with space. And while I'm pretty sure Edwin Hubble loved nothing more than looking at space through a telescope, I can't help but wonder if he would have wanted to go there in person too.

Q4: HOW DID AMERICA ENTER THE SPACE RACE?

MY PARENTS WERE IN college in Taiwan when America's peaceful journey to space began after World War II, one of the most violent periods in human history. While I later watched awestruck as Apollo astronauts and Soviet cosmonauts suited up to make history, at the time, I learned little about the men and women who did the math and built the rockets and vehicles that carried astronauts into space. When I got older, historical figures emerged that resonated with me. They were not mainstream famous, but just as important to space exploration as the celebrated Moonwalkers—maybe even more so. They were the engineers.

The laws of physics and chemistry gave early aerospace engineers a run for their money. There were two problems to solve if we wanted to fly into space. One, creating a way to get out of the Earth's atmosphere. The other, building a spacecraft to orbit the Earth and then come home. The answers came from a collective of brainpower from across the globe.

One of those brains was from my parents' beloved China—an engineer named Qian Xuesen who was instrumental in designing the rockets that sent Americans to space.

As I mentioned, my parents never knew a peaceful China. At dinner when we were kids, my parents sometimes talked about how bad the war was, how my grandparents barely survived. The ruling Nationalist Party had been fighting a civil war since 1927 against communist guerrillas. And beginning in 1932, Japan sent forces to attack Shanghai—dropping bombs with airpower the Chinese had never seen.

Qian Xuesen was a college student who lived through that first Japanese bombardment that killed thousands of his countrymen. At the time of the attack on Shanghai, China owned only ninety planes that could fly (Japan had 2000 planes and a growing aviation industry). The pathetic Chinese "fleet" was divided up among warring generals with zero ability to deploy them with any skill. Qian knew that it wasn't just important to create new technologies for social progress, it was a matter of national defense. After graduation, Qian was hoping to learn about aeronautics, but China didn't have a single graduate program for aviation engineering. After centuries of thinking more and building less, China had accomplished little in the way of modern munitions or aircraft.

Qian was super smart, though, and he landed a US-government-funded graduate scholarship at the Massachusetts Institute of Technology with the requirement he would return to China with some helpful skills. The idea on America's part was to get our Eastern ally up to speed with technological advances. It was clear China was getting pummeled by their highly advanced neighbor, Japan—our growing adversary.

The scholarship required Qian to travel around China for a year and get a read on what their current situation looked like. There was little to see. He toured dusty hangers with a few junker planes and some rusty engines lying around. There were hardly any mechanics who could repair them. Near Shanghai, there were some American Curtiss Hawk biplanes in mud bunkers and not much else.

An interesting aside: Around that same time, my great-uncle on my mother's side, *Xie Chun Sung*, was training to be a pilot in China during the early days of the war. He trained with the American Volunteer Group called the Flying Tigers, which was headed by General Claire Chennault. Then he was sent to the US for advanced flight training, likely in Arizona and Idaho. He transitioned to bombers and the F-5E photo reconnaissance version of the Lockheed P-38. He survived all his combat missions during the war, only to be killed later in 1949 while still flying in *Chang Kai-shek*'s air force in Taiwan. He is the only relative of mine that I know of who also loved to fly.

Throughout 1934, the civil war raged, and Japan kept up their attacks. Without much to do scientifically, Qian filled his brain with live performances of the Shanghai Symphony Orchestra conducted by Italian pianist Mario Paci. To Qian, the music was not only an auditory delicacy. He visualized the notes as colors. In 1935, Qian stepped aboard the USS *President Jackson* in Shanghai and sailed for America.

The Massachusetts Institute of Technology turned out to be a difficult fit for Qian. An aristocrat whose mother's feet were bound (small feet were considered beautiful and conferred great social status), Qian spent his childhood studying science and making calculations instead of tinkering and building. In China, thinkers and doers were separated by class, and rarely did the two mix. In America, technical schools wanted students and researchers that could do both. This was a challenge for Qian. "Americans are born with monkey wrenches in their hands," Qian lamented after watching fellow grad students create experimental flying machines and test engines. Qian got stellar grades and worked on a thesis about wind effects on airplane wings, but he could not figure out how to turn his theories into practical experiments. And he didn't want to.

So he left for California to have a meeting with the famous Jewish Hungarian aerodynamicist, Theodore von Kármán. Von Kármán, a charming and impulsive bachelor who ran the Guggenheim Aeronautical Laboratory at the California Institute of Technology, was delighted and extremely impressed by the sophisticated young Chinese theorist in the fitted suit and silk tie. He didn't care that Qian wasn't interested in experiments. The two bonded immediately. Qian became von Kármán's translator. Whatever von Kármán's overactive brain could imagine, Qian transcribed into beautiful numbers and formulas on paper. As close as the two became personally, Qian always referred to von Kármán as "My revered teacher." For von Kármán, Qian was the most gifted student he'd ever had—and the Guggenheim lab was full of brilliant graduate students.

Qian worked on compressible fluids for rocket fuel and formulas for subsonic flight. In his spare time, he met with a rocket club started by fellow graduate student Frank Malina. The four other students welcomed Qian as their "resident mathematician," and he welcomed a new challenge: *What should a nozzle's angle be? What is the best way to create thrust? Chamber temperature? The ideal rocket motor?* He worked it all out with a pencil and paper. The rocket club convinced von Kármán to give them basement space in Cal Tech's Guggenheim lab. But after a series of serious mishaps where they blew up a bunch of lab equipment, the five students got monikered the *Suicide Squad*, and the university kicked the club out of the building.

The squad took their incendiary experiments to a nearby dry riverbed. They did so much good work with a thousand-dollar grant from a fellow student that they got the Department of Defense's interest. Now with a hefty government budget, they were in the big time. Von Kármán wanted to shake the stigma that a "rocketry club" was a silly hobby for eccentric millionaires and thrill seekers. Qian, Malina, and von Kármán got together and came up with the name, Jet Propulsion. They crowned their new endeavor the Jet Propulsion Laboratory and everyone called it JPL.

In 1945, at the end of World War II, US Army General Henry Arnold called von Kármán to a secret meeting. Find the German rocket scientists, Arnold said. Figure out what they know and convince them to come back here before they take another offer. Arnold had a new vision for the US military. "For twenty years," Arnold said, "the Air Force was built around pilots, pilots, and more pilots. The next twenty years is going to be built around scientists." This was pure poetry to the JPL teams.

Von Kármán knew that he had to take Qian with him. Who better to quickly size up what marvels were going on in Hitler's facilities that were now fully under our control? We had to get the good people and equipment before the Soviets did. Together in May of 1945, the Hungarian American and the Chinese national dressed in US Army uniforms to embark on the awkwardly named Operation *Lusty*. They carried lipstick and stockings in their bags as "gifts for blondes"—a cover of US officers looking for romance. For sure what Qian found piqued his interest far more than any lusty *fräulein*. One of the very first German scientists Qian interviewed was the famous German rocket designer Wernher von Braun, who had surrendered to American forces after his laboratory was captured by the Soviets.

When I began working on a project with the Jet Propulsion Laboratory as an employee of Hexcel Corporation after finishing my PhD in 1987, I'd never heard of Qian Xuesen. Later, on a visit back to Lawrence Livermore after starting my astronaut training, a Chinese engineer handed me a book by an American writer named Iris Chang: *Thread of the Silkworm*. I was amazed. *A Chinese aerospace engineer helped create America's top rocket research facility?* Here was a young scientist from my parents' home country, from a time when China had almost zero technical players on the global stage. And he became a top contributor to American rocket dominance. *Wow.*

I wish this story had a happy ending for our country, but McCarthy politics and the "red scare" crept up on Qian as Mao Zedong's communist party took power in 1949. In 1950, the FBI raided Qian's house (found nothing), canceled his security clearance leaving him unable to work, and put him under house arrest for five years. Von Kármán and top academics wrote letters begging for his release—to no avail.

The United States government stripped Qian of all his notes and papers and bought him a one-way ticket home. He arrived in China in 1955 with his brain and memories intact, into the arms of Chairman Mao. Qian launched China's missile and space programs and designed the "Long-March" rockets and "Silkworm" missiles that would later be fired against the United States forces fighting for South Vietnam.

In the end, there would be no Chinese aerospace engineer in the American rocket program headed by von Braun, the former chief rocket engineer for the Third Reich.

PART TWO

APPETIZERS

Flavorful Inventions and Players in the Race to Space

Q5: HOW DID THE SPACE RACE BETWEEN THE US AND THE SOVIET UNION START?

TWO BRILLIANT MINDS IN space history began their careers like I did, cutting and gluing balsa wood to glide on the wind. American friends Max Faget and Caldwell "CC" Johnson turned their love of model airplanes into machines that sat atop rockets and carried humans into space. While Qian Xuesen was making calculations in California, on a strip of sand called Wallops Island in Southern Virginia, Faget and Johnson were having the time of their lives exploding things too.

"After World War II there were many surplus solid propellant rockets around," Johnson said in a later interview. "And this was a whole new technique of getting data; and it was wide open. You could do anything you wanted."

What they wanted was to see how fast, how far, and how high they could go.

Since the time both were kids, Faget and Johnson designed and built model airplanes. They learned to understand pressure, lift, and the friction the wind had on an airplane body. They'd even been at the same model airplane competition when they were teens. Faget was an incessant idea guy, absent-minded in the ways of schedules, timing, and manners. To him, everything was a toy. There were machines that flew, submarines that dove under the water, cars that raced along tracks, and rockets that shot into the sky.

Johnson was the reality to Faget's wild ideas. Like Theodore von Kármán and Qian Xuesen, Faget and Johnson worked in a high-level symbiosis. Colleagues would say they transferred thoughts through osmosis. Johnson put Faget's dreams on paper and sketched out all the necessary components to make them work. The two had the utmost admiration for each other. "Caldwell knows what the shops can build, and how well it will hold together," Faget said in a 1991 interview. "And this is infinitely more valuable than designing something that may look like a masterpiece but will fall apart."

Among the swamp nettles and bloodthirsty mosquitoes at Wallops Island, Faget, Johnson, and a handful of other engineers (most without college degrees but many with model airplane–making know-how) slept in tents with wooden floors near the cement launch pad. Their paychecks came from a government agency called the National Advisory Committee on Aeronautics. Their mandate was this: Take the military leftover rockets from WWII and do all the tests you want. Reconfigure things to make them go higher and faster. Explore the speed of sound and beyond. For Faget and Johnson, it was paradise.

One day the engineers were sitting around at lunchtime playing cards. Johnson remembers, "People started saying,

'Well, if we just made it a little bit bigger and had a few more things, we could put a man in it.' And I think that's what started the whole business of the space program."

Unfortunately or fortunately for the American model airplane makers turned rocket scientists, on October 4, 1957, the Soviet Union launched the first Earth-orbiting satellite, Sputnik 1. Then a month later, Sputnik 2 carried a dog named Laika into space.

America was caught off guard. President Dwight D. Eisenhower downplayed the accomplishment, saying the that the Soviet feat was "one small ball in the air." After blowback that he was not taking the Soviet national security threat seriously, Eisenhower started allocating funding to play catch-up.

CC Johnson was worried that the space program would end up as part of the military like it was in the Soviet Union. "But the President finally decided it was going to be a civilian outfit," he remembers. "And NASA was born." It was a bit of a shock for the men in the marsh. "You know, it's one thing to sit around the table at noontime and play Pinochle and bullshit," Johnson said. "It's another thing for the President of the United States to all of a sudden tell the world what you're bullshitting about!" It was, Johnson said, time to "put your money where your mouth is." They were young and undaunted.

Back then there was a sharp divide in the NASA ranks: *To man or not to man.* Things got heated.

"Why should we waste 150 pounds and learn about all the safety of a man when with a quarter of that weight you could automate the system?" the automation guys argued.

"Well, what's the sense of doing all of this if a man is not involved?" the manned spaceflight team countered.

Max Faget worked on a design for a space capsule NASA called Mercury, after a winged Roman god. Johnson loved the simplicity of the machine, a metal cone with a blunt top. Fortunately for me and my astronaut colleagues, Johnson configured an astronaut into the Mercury design because this was where the space race was headed, and this was what got people excited. To fit America's first astronaut, Alan Shepard, into the claustrophobic space, Johnson thought of the man as a physical object. "I didn't design life support systems. I was interested in the physiology of the man simply because it was a mechanical system as far as I was concerned, and he had to be supported mechanically. There are some things that men do wonderfully, and some things they're not very good at. The instruments can do things quick, but some things they can't do worth a damn. They can't think, for one thing."

Johnson understood forces and loads—how much a body could take in what type of situation. How could a conglomerate of blood, flesh, and bones survive being shot up on a rocket, orbit the Earth, and come down without burning up or smashing to smithereens? He sketched and calculated and rearranged tubes and wires. He discovered that the astronaut needed to lie on his back across the diameter of the spacecraft. "It's hard to fit a man into the little thing that's got a point," he said. He found that Shepard would need to lie on his back—"eyeballs in"—because it was known that would put less stress on his corneas under high G-forces. It was also a challenge to add a window or a way to see out without losing structural integrity. On this point, Johnson argued that human eyes were, "the best instruments there ever were," and the astronauts needed to be able to see outside. *Thank goodness.*

On May 5, 1961, Alan Shepard went into suborbital flight in a six-foot-wide spacecraft named *Freedom 7* and became

America's first astronaut. We were still behind in the space race, though, because it was almost a month after Yuri Gagarin, the first Soviet cosmonaut, flew in space. Then the Soviets beat us with the first walk in space. America was not used to coming in second. This was getting annoying.

Up next for Faget and Johnson was the storied Apollo program. It was 1962. President Eisenhower was out, and the new flashy Democratic president, John F. Kennedy, wanted men on the Moon, *stat*. NASA was budget-fat, and with so much money and contractors vying for pieces of the game, things got complicated, and then tragic.

There were a lot of bugs in Apollo 1, including the communication system, which glitched in and out. Sadly, there were worse problems. Ed White—already a national hero after completing the first American spacewalk (just *after* the Soviet spacewalk, *ugh.*)—joined fellow astronauts, Gus Grissom and Roger Chaffee, on a mission to check out the Apollo capsule in preparation for future Moon flights. The plan was to orbit the Earth a few times to get a read on how Apollo performed out there. A month ahead of the February launch, the three men climbed into the capsule on top of an unfueled Saturn IB rocket. They closed the hatch, and the capsule was filled to pressure with pure oxygen.

As the men lay on their backs testing the machinery, faulty wires sparked a fire. Garbled communications came through to the Mission Control Center: *Fire! I smell fire! Fire in the cockpit!* The last transmission was an agonizing cry. NASA investigators believed White tried to open the hatch, but because of the pressure in the capsule working against it, he couldn't get it free.

Caldwell Johnson was devastated. In space, the capsule would be at low pressure, but at sea level, "it was a bomb ready to go off," he said. NASA investigators blamed North American Aviation, the contractor that built and wired the capsule. Johnson disagreed. "A bunch of us were all involved in it. And we all had better sense. I think all of us that had a place in the decision to use that system should take the blame. Not any one person, but collectively, we should have known better. It was a terrible mistake. You can't undo it."

At the time of the Apollo 1 tragedy, I was six. I was obsessed with airplanes, and the idea of a spaceman career was taking hold. I saw the charred remains of the Apollo 1 capsule in the newspaper the next morning and read that the three astronauts were dead.

"What happened?" I asked my parents over dinner. "Why was there a fire?"

"Not sure, son," my dad said, "It was an accident." My parents didn't say much else. It was sad, but I was a kid. Death was abstract. Space was real. I wanted to see it in person.

Faget and Johnson kept working on bettering the Apollo along with ever-growing teams designing the rockets, the capsule, and the Lunar Lander. But their ideas were way ahead of Apollo. Even before Neil Armstrong started training for the historic Apollo 11 mission, Faget was in his office in Houston cutting and gluing a paper model of what he thought should be next on NASA's to-do list: a reusable spaceplane that would shuttle astronauts to and from space—a paper prototype of what I would one day ride into orbit three times.

Q6: WHO WERE THE FIRST SPACE WALKERS?

I DON'T KNOW WHO first dreamed of the possibility of walking in space, but my hat goes off to them. With the help of genius engineers who understood the laws of physics in a vacuum and zero-g, the first human stepped into the void of space more than half a century ago and launched one of the most elite clubs in human history.

Even though the Soviet space agency had tight budgets under communism, once Yuri Gagarin flew out of the Earth's atmosphere and circled the globe in 1961, Soviet aerospace engineers let their imaginations go wild. With the idea of getting a man outside a spacecraft, they designed a clever airlock that inflated from the round belly of their Voskhod 2 spacecraft like a tunnel on a kid's bouncy house. And to keep the spacewalker alive in zero atmosphere, no gravity, and temperature swings around minus- and plus-250 degrees Fahrenheit, they designed and built a pressurized spacesuit.

Voskhod 2 launched in March of 1965. Once it was safely in orbit, the crew depressurized the cabin, and cosmonaut Alexei Leonov emerged from the airlock headfirst into space. He was tethered to the ship by a five meter "umbilical cord" that supplied oxygen and kept him from floating away. Leonov was mesmerized. "I saw at once the Black Sea and the Crimea Peninsula, then I turned my head to the left and saw Romania, Bulgaria and on the end I saw the boot of Italy. By raising my head, I saw the Baltic Sea and the Kaliningrad bay, familiar to me from childhood. Stars were to my left, right, above and below me. The light of the sun, was very intense, and I felt its warmth on the part of my face that was not protected by a filter. The sky was black and the silence was extraordinary." Pavel Belyayev was Leonov's only crewmate and the commander of Voskhod 2. The rumor was that the Soviet space program chief had told Belyayev that if for whatever reason Leonov can't get back in, and he's basically going to die, take this pistol and shoot him. Put him out of his misery.

So, you can imagine Leonov's heart pounding when after twelve minutes of pure joy, he went to head back in and found his suit over-pressurized like the Pillsbury Doughboy. His hands slipped out of the gloves so he couldn't grab anything. In the vacuum of space, he struggled and struggled. He sloshed in his own sweat. Tick . . . tick . . . tick . . . He thought about going in headfirst, but the capsule was so tight there would be no way to turn around. He shoved his ballooned legs as far as he could, but his backpack caught outside the edge of the airlock.

Suddenly, Leonov had an idea. He turned down the pressure regulator on his suit until it began to deflate. It was a dangerous move. Dropping the pressure could force nitrogen from his cells to bubble into his blood and kill him. And the lower oxygen may not be enough to keep him alive. *But what do I have to lose?* he thought. He deflated until he was flexible enough to bend. And with just seconds to go before things turned critical, he slid back through the airlock into his copilot seat and the cosmonauts took off for home.

Leonov and Belyayev parachuted back to Earth hundreds of miles off course. Voskhod 2 rolled down a snowy mountain almost over a cliff. Luckily, the parachute caught a tree and kept them from going over. Wolves howled as they tried to keep a fire going in a snowstorm until rescue teams arrived two days later.

Around the time Leonov became the first human to walk in space, I was in first grade, and also madly in love with my teacher at Goddard Elementary school in Wichita, Kansas. Miss Link was *perfect.* I asked my parents if we could invite Miss Link for dinner, and I watched, dreamy eyed, from across the kitchen table, as my midwestern teacher enjoyed Mom's Chinese cooking (Mom set Miss Link's place with a fork instead of chopsticks, of course).

In those years, the Cold War played out between the US and the Soviet Union in the movies, in the space race, and in our daily lives. There was an airbase near my hometown, and so every now and then a sonic boom rattled my bicycle handlebars and vibrated my chest. I was highly competitive and very patriotic. I felt a stab of envy when Alexei Leonov, wearing his *Soviet* spacesuit with a red flag patch, opened the hatch, crawled out of the ship, and became the first man to walk in space.

"Why are they always beating us?" I grumbled to my dad after school. He ignored me.

My parents, especially my dad, were always neutral when it came to people from different places, times, and experiences. On the other hand, I was very much a product of American propaganda and was told repeatedly that the Soviets are evil. Still, I was impressed with their space skills.

Just three months after Leonov's historical spacewalk, our country watched Ed White become the first American to walk in space—what NASA dubbed conducting "Extravehicular Activity" or EVA for short.

White's mission was less chaotic than Leonov's but still a fantastic adventure that had me glued to the radio and TV for updates. White was an Air Force fighter pilot from Texas chosen for the Gemini IV mission. It wasn't guaranteed that he would get to do a spacewalk, but the Soviets had just trumped us, so NASA made sure White's spacesuit was ready in case they could make it happen. White was raring to go. He hedged his bets with three religious medallions he carried with him: a Saint Christopher medal, a gold cross, and a Star of David. They may have come in handy because there was a faulty latch on the hatch. So once White went out, they might not have been able to close it securely enough to reenter the atmosphere without burning up.

The television feed of White floating in slow somersaults—inflated legs and arms spread like a kid's toy against a big blue Earth—was awe inspiring to my five-year-old self. Unlike Leonov, White had some control over his moves. He used a nitrogen-powered "zip gun" to jet playfully around—clearly having the time of his life. "This is the greatest experience, it's just tremendous," he radioed to Houston. "Right now, I'm standing on my head and I'm looking right down, and it looks like we're coming up on the coast of California. There is absolutely no disorientation associated with it."

White traveled from Hawaii to Florida in fifteen minutes, zooming along at over 17,500 mph. And like how I hated to quit playing astronaut when my parents came home from work, White didn't want to come back in the capsule. Then, like my mom at dinnertime, Houston called and demanded he get back to the ship. "I'm coming back in . . ." he radioed, "and it's the saddest moment of my life." I felt his pain.

Q7: WHEN DID THE SOVIETS AND AMERICANS START WORKING TOGETHER?

AFTER MONTHS OF PRACTICING our lunar landing in my folk's garage, on July 20, 1969, my best friend from school, Russ, and my neighborhood buddies Jack and Pete came over to watch the Apollo astronauts take *their* Moon shot. My dad never seemed overly excited about anything, but this was a big enough event for him to move our old black-and-white television (with rabbit ear antennas) outside so we could watch the landing in our backyard. It was almost ninety degrees, and we had no air conditioner.

Dad made peanut butter and Spam sandwiches for lunch (they're actually not bad). Mom cut juicy watermelon slices that dripped down our chins, and we competed to see who could spit the seeds the farthest into the yard. In the afternoon, we sat on lawn chairs in the shade glued to the TV. We watched the mission control team and listened to the radio calls back and forth with the men in the Apollo capsule and lunar module. The engineers were Dad's age and wore short-sleeve white shirts and headsets. They looked nervous. I gripped the plastic armrests on my chair with sticky hands and bit my lower lip. Then we heard the words:

The Eagle has landed.

We jumped to our feet along with the men on TV.

"Wow," Jack said, jumping up and down. "They're on the Moon!"

"Yes! They landed!" I yelled. "They landed on the Moon!"

My dad was quiet and contemplative. I was pretty sure he thought it was cool too.

When it got darker that evening, I couldn't take my eyes off that glowing light in the sky, almost a quarter of a million miles away. *Neil Armstrong and Buzz Aldrin are up there getting ready to walk on the surface of the Moon. I want to be like them. That's what I want to do. I want to be like them.*

With the Moon landing under America's belt, things cooled down between NASA and the Soviet Space Program. We were the clear winners. Everyone at school and in the neighborhood was so excited at what had been accomplished. We deserved the bragging rights.

I was in sixth grade in 1972 when the US made a deal with Leonid Brezhnev to collaborate on some kind of space program. The two heads of state thought it made sense to team up for a symbolic act of friendship since our two superpowers were leaps ahead of the rest of the world. A lot of ideas were thrown about, and both sides agreed to rendezvous a Soviet Soyuz spacecraft and an American Apollo capsule in orbit. (The word *Soyuz* means "union.") Back in Houston, CC Johnson went to his desk to sketch out a way to dock the two spacecraft that had been designed six thousand miles and cultural worlds apart.

Q: DID YOU THINK IT WAS A COOL IDEA?

TO A LOT OF AMERICANS, the Soviet-US space collaboration was suspect. To my sixth-grade brain, it was just wild. From everything I'd learned up until now, "evil communists" were planning to come and take over America. When I was in kindergarten, we lived in Wichita, Kansas, where an air base nearby blasted sirens anytime of day or night. Random sonic booms exploded in my ears. In school, we readied for doom and crawled under our desks to practice surviving a nuclear bomb.

I got most of my news from the car radio when Dad drove us places in our unsafe-at-any-speed Corvair and later Karmann Ghia. I heard the American astronauts were learning Russian. The cosmonauts were coming to Houston to train. I was suspicious, but it was space exploration with astronauts and cosmonauts, so I couldn't help thinking it was cool. To me, the Soyuz spacecraft looked goofy—like a giant bug with awkward rectangular wings (actually solar arrays) and antennae sticking out from all sorts of places. The Soviet craft didn't even need pilots, because you could fly it from the ground by remote. *How lame.* Our Apollo capsules were smooth, conical—nice and clean. We were so much better.

When it was "go" time for the docking of one capsule to

another, my love for space overcame my hate for "evil communists." Live on television, I watched the two spaceships join together without a hitch thanks to the design of CC Johnson and his team who worked with the Soviet engineers. When the hatch opened, NASA astronauts Thomas Stafford, Vance Brand, and Donald "Deke" Slayton, reached through and shook hands with cosmonauts Alexei Leonov and Valeriy Kubasov. It was cool. But still, I didn't trust those guys.

Q8: HOW DID ROCKETS FIRST GET TO SPACE?

FOR ME, THE STORY of how rockets got to space is multi-chaptered. To go to space, humans first needed to fly. To fly, humans needed to understand how birds do it. How does a wing create low pressure above to lift? How does a body propel forward through the air? How can feathers help to steer and change direction? These questions were my middle school kid obsessions. I needed to know how to make my paper airplanes glide and how to build my models. I rode my bike to the public library and found every book I could on how things fly.

I copied airplane drawings in the books onto tracing paper. Then I took carbon paper and traced them onto balsa wood. I cut out the wings, bodies, and tails. I made little models of the different airplanes. One of my favorite books in the library was a small hardcover one that featured military airplanes during the early days of jet fighters and bombers. On one page was a brief description of the airplane, and on the facing page was a three-view of each one in profile, top, and front views. Two of my favorites from that book were the Douglas F4D Skyray and the Avro Vulcan. They were sweet looking, and both featured large delta wings that made them glide particularly well.

My balsa wood planes were gliders or powered by rubber bands that I attached from tail to propeller and wound up as tightly as I could. Trial and Error were my friends. I didn't fully understand the physics behind it, but I knew certain rules of thumb. I added weight to the nose to get the plane's center of gravity about one-third the way down the width of the wing. I learned about control surfaces: ailerons (to roll side to side), the elevator (to move up and down), and the rudder (to steer and make coordinated turns with the ailerons). My models went from nose diving into Mom's flower garden to flying halfway down the street. I was ready for bigger things.

One library book told the story of an aerospace engineer named Francis Rogallo. In the 1950s, Rogallo worked at NASA's Langley Research Center in Virginia where Max Faget and Caldwell Johnson were rushing to beat the Soviet Space Agency. Collaborating with his wife, Gertrude, the couple came up with a triangle-shaped, inflatable "delta" wing with two scalloped lobes that generated lift. The Rogallo wing was a novel design because the fabric shifted with the wind more like a boat sail than a hard airplane wing. Rogallo pitched NASA on the idea that his wing could unfurl, inflate, and glide the Gemini capsule as it reentered the atmosphere to a splashdown. In 1961, NASA renamed it the Parawing and contracted North American Aviation to build them and run a bunch of experiments. After problems with the wing falling apart and the amount of effort it took to pilot the wing from inside the capsule, NASA ditched the project. The Gemini spacecraft and those that followed would come down by parachute.

Rogallo was bummed, but he and Gertrude loved their design and the idea of a personal flight craft. They spent many hours after work fine-tuning the wing. They trusted their own know-how enough that, along with their four kids, the family acted as test pilots. On windy North Carolina beaches, they strapped in under the wing to jump into the air from the dunes and glide above the shore. The Rogallo wing caught the attention of thrill-seeking Americans who wanted to fly. The couple had a patent on their design, but as others tweaked it and sold new models, rather than sue, the Rogallos became the beloved "parents" of the hang-gliding community around the world.

I thought the Rogallo wing was awesome, and I intuitively understood the aerodynamics that made it work. Our housing development was still under construction, so after school and once the workers left for the day, I combed the sites and found an eight-foot-long one-by-two-inch piece of pine. I felt guilty because it clearly wasn't scrap wood like I normally collected to build model boats and cars. But I couldn't resist. I carried it back to our garage and spent weeks sawing it lengthwise and making a cross to form a big triangle. I took a roll of plastic that Mom had for her vegetable garden and nailed tiny brads through the plastic to the wood. Below the wing, I added a crossbar to hold on to. I used speaker wire as a frame to constrain the plastic from the top of the triangle down, and across the base (looking like a triangle with a line down the center from top to bottom) so that it would form the Rogallo shape when inflated. I was quite proud of myself. I had my own Rogallo wing, and I would fly.

At the construction sites, men on big earthmovers graded different levels to pour foundations for future homes. It was the weekend, so I had the place to myself. I scouted until I found the best berm, about five feet high. I got under the wing and held the crossbar. The wind grabbed my sail, and I heard the plastic *flap flap flap* against the breeze. I expected to take off like the guys I'd seen on extreme sports TV. I walked back about twenty feet to get a running start . . . run run run . . . jump and . . . *Bam!* I wiped the dirt from my shirt and rubbed my aching knee through my dirty jeans.

Okay, not bad. I think I need more speed.

My wing wasn't damaged, so I hiked back up to the launch site for my second flight test.

I started running from a bit farther back. I jumped off the berm. I felt a few seconds of glide and then . . . *Thunk!* I was down. Same results after a few more crashes. I knew enough about physics to know the problem. The payload (me) was too heavy for the lift (the upward force produced by the higher air pressure beneath the wing relative to the pressure above the wing) that my Rogallo wing could produce.

I sat in the dirt and thought and thought. Typically, scientists and engineers don't cry or get discouraged when our experiments flop. Failure is a way to collect data and the impetus to come up with your next strategy.

Aha! Yes!

My plan B was back home, likely playing with her stuffed animal collection or coloring with her crayons.

"*Xiǎo Pei!*" I yelled, barging into the family room.

My younger sister, Sandy, looked quizzically at my interruption. It wasn't often I paid attention to her.

"Listen, I need you to help me with an important experiment."

At this point in our lives, Sandy still had a little bit of interest in pleasing her big brother.

Back on the berm with a new payload (she was six years old and weighed about half what I did), I positioned Sandy under the wing.

"Okay, *Xiǎo Pei.* Hold on. And whatever happens, don't let go."

I ran alongside her and pushed her off to give her the thrust she needed.

"Gēgē!! Big Brother!!" she wailed.

For a few seconds, I gathered flight data, unbothered by her screams.

Then my little sister crash-landed and started crying. I was unfazed. She wasn't *really* hurt. I'd gotten bruised up a little bit when I jumped, so maybe she'd have a few bruises too, but no biggie. I knew I wasn't the nicest brother, but this was important work.

"Come on, you're okay, *Xiǎo Pei,*" I said, annoyed. "Wasn't that fun? You were flying!"

"Owwwww," she went on.

It looked like I wasn't going to get her to do a second test run, so we walked back home, my mind spinning an explanation as to why my sister was sobbing, snot-nosed, and covered in dirt.

"*Érzi*, what happened? What did you do to your sister?" Mom demanded.

I mumbled, "Uh, she . . . fell?

"He . . . sob . . . made . . . sob . . . me . . . sob . . . fly . . ." Sandy wailed.

"*Érzi,* that is dangerous," Mom said. "You shouldn't do things like that to your sister."

Historical note: It's been over fifty years. Sandy still brings this up at times.

My Rogallo wing was too small to carry even a six-year-old. But it gained the interest of my friend's engineer dad down the street. He inspected my wing with some admiration and came back with heavy string. "It's not big enough to carry a person," he said, "but I bet it will make a great kite." He attached the string and started running down the street with it. Sure enough, it flew wonderfully in the wind.

Q: AND AFTER YOU UNDERSTOOD THE PHYSICS OF FLIGHT?

AS ASTRONAUTS STARTED GOING to space, I switched my focus from airplanes to rockets. Getting a person to fly through air is one thing, but if you want to go to space, that's a whole other thing in the world of physics. In space, there is no air to give wings lift or create thrust with air-breathing jet engines. *So how do we get there?*

Sixth-grade summer school meant I could take fun or interesting classes, so I chose a class on rocket building. On the first day, our teacher gave us each a model rocket kit. We learned how to cut fins out of balsa wood, how to glue them to a paper tube, and how to stabilize the rocket with weights in the nose. We added solid rocket motors to the base. This is where chemistry came in. Up to now, my experiments with planes and gliders only dealt with physics. To get a rocket to blast off, you needed chemistry too. In class, making rocket motors from scratch wasn't an option (remember the Suicide Squad at JPL blowing up the lab?), so we bought them from the hobby shop.

I wasn't sure what was in the solid rocket motors that we bought, but I knew that it worked in a similar way as gunpowder. I also knew, if I wanted to understand rockets and explosives outside of summer school, I'd need to learn some chemistry. Fortunately, there was an article about gunpowder in one of my favorite places—our set of encyclopedias my parents bought from a door-to-door salesman. The article detailed the chemicals I needed and what the relative proportions should be. Like a good scientist, I decided that I would make gunpowder as an experiment.

The fuel for gunpowder and rocket motors is carbon, just like the charcoal that you light to grill your hamburgers. Charcoal is a dense concentration of carbon atoms that burn very hot. I took a few briquettes from the bag in the garage and smacked them with a hammer. It was surprisingly tough to break apart, but I eventually pulverized it. Then I added sulfur—the stinky yellow element on matches—as a secondary fuel and a catalyst to help the charcoal burn. Then I needed need an oxidizer. This is the substance that supplies oxygen to keep the combustion going—like how oxygen in the air keeps a forest fire burning. For gunpowder, the oxidizer is usually potassium nitrate (saltpeter), a compound that has three oxygen atoms on every molecule. Fortunately, I could get sulfur and potassium nitrate at the local drug store. I didn't even need an adult with me. I simply asked the kind pharmacist, and he sold them to me with a smile.

I wasn't making gunpowder to make my own rocket engines. I knew that gunpowder burned a bit too quickly to be suitable for rocket motors. There was a reason that it was used in rifle and pistol cartridges. Instead, like a lot of boys, I wanted to make my own bombs and fireworks. In addition to the necessary chemicals, I bought fusing by the foot from the local hobby shop. (It seems amazing that what was available and sold to enterprising kids without question back then would get you on a no-fly list today.) My gunpowder worked, and I made a few small bombs using old model paint bottles. My friends and I blew them off in the street next to my house. Sadly, or maybe fortunately, my chemical-engineer parents got wind of what I was doing and put a stop to it.

"What are you doing?"

"Uh . . ." I was dumb enough to tell them the truth.

"Where did you get the chemicals? You're going to hurt yourself! No more!" Live and learn.

I went back to model rockets and store-bought rocket engines. They were what we call "solid" rocket engines. A rocket's motor is not like a car engine. A car motor burns fuel to create gasses that pressure mechanical parts and make you move. The rocket uses a "chemical" motor where the ignited fuel creates a force called thrust by releasing hot gases as the propellant burns. Physics comes back into play with Newton's third law: *For every action there is an equal and opposite reaction*. So, the thrust of gas pushes against the combined forces of gravity and atmospheric drag that try to keep the rocket on Earth. Whichever force is greater, that's the direction that your rocket will go.

After the summer class was over, I built rockets nonstop. Setting them off was another thing. In California, the laws for rockets and fireworks are super strict because there is always a burn hazard. You can only launch rockets with an adult that has a license and permission. I *needed* to launch my rockets. So, in seventh grade, I convinced the science teacher to be the adult in charge, and I started a rocket club. We met after school to build our rockets, and on certain Saturdays, we'd go out to the school's field, set up launch pads, and fire off our rockets. It was awesome.

"Ok, Leroy, you're next."

Yes!

I set up a rocket that was my own design. I bought the necessary parts from the hobby shop. I built my vision of a single-stage-to-orbit (SSTO) spaceplane vehicle that featured ramjet engines (a novel jet engine design with minimal moving parts) mounted above the body of the craft. From my reading, I knew that rocket engines were more efficient up high, so I thought that using ramjets in the lower atmosphere to help gain altitude quickly might make sense (although, of course, the "ramjets" on my model rocket were not functional). I did my countdown and pushed the ignition button.

She flew like a dream.

"Bravo, Leroy," the science teacher said. "Well done." The other kids praised my design too.

I beamed.

Q: WHY WERE YOU SO INTO ROCKETRY?

YOU COULD SAY I had rocketry in my blood. It turns out that my parents, and by extension, I, came from a country with a long line of chemists. Way back in China during the Tang Dynasty (9th century), Taoists monks in elaborate robes isolated elements and compounds to create magic potions for long and/or eternal life. When the alchemists mixed sodium nitrate, charcoal, and sulfur, a big burst of gas exploded from the vessels and sent shrapnel flying in all directions.

The monks didn't create immortality, but it was the beginning of a fireworks bonanza in China that continues today. Chinese love their fireworks (me included, as evidenced by my short bomb-making career), and the brilliant displays are part of every big celebration. The following Song dynasty had another idea of how to use the exploding concoction. Rather than finding eternal life or creating art, *maybe it would be a good way to take out our enemies.* Song rulers had a robust military industrial complex where contracted inventors worked on best ways to use the explosive powder. They created a very active *huǒjiàn* (literally, fire arrow) program and launched their *huǒjiàn* against Mongol invaders in 1232 at the battle of Kai-Keng. The *huǒjiàn* weren't fancy—just a piece of bamboo with gunpower packed in one end, open on the other, and attached to a longer stick for stable flight.

The Mongols quickly caught on to this flashy form of warfare and started building and improving on the technology. They first shot their rockets into Baghdad in 1258, just twenty-six years after being on the offensive from China. The rockets were rudimentary and had little guidance capability, but the Mongols were extremely brutal and laid the intellectual city of Baghdad, known as *Madīnat as-Salām* (the City of Peace), to waste.

Twenty-five warring years later saw the Mongols consolidate much of Asia and set their sights on a nearby island known to be full of riches. The Mongols attacked Japan with fireballs that exploded on impact. (*Note:* Some historians say they used rockets too. As with a lot of ancient history, some historians say they didn't.) The Japanese and their samurais were not having it. They had a whole different way of fighting than the Mongols and a superior defense that turned to offense and pushed the Mongols back to sea.

Soon, rocket and other chemical explosive technologies were adopted by countries around the world who clearly saw that humans were going to need bigger and better ways to kill each other for the foreseeable future.

Q: BUT WHEN DID MODERN ROCKETS COME ABOUT?

MY PARENTS SUBSCRIBED TO a series of wonderful books published by *Life* magazine on a variety of topics. My favorite, of course, was the one on rockets and space exploration. I read all about the work of rocket scientist American Robert Goddard, parallel to the work of Wernher von Braun in Nazi Germany.

Goddard consulted for the US Army during the Roosevelt administration, but most of his efforts were focused on reaching space. Von Braun also dreamed of spaceflight, but he knew that creating weapons of war was where the funding was in Nazi Germany. In the cutaway drawings of the rockets in the *Life* book, it struck my child's mind that the payload on Goddard's rocket was a parachute—on von Braun's, it was high-explosives.

I didn't know about the Holocaust until I was a little older. And even when I did, I didn't learn about the mammoth, subterranean chamber of horrors where Nazis used slave labor to build von Braun's rockets to destroy parts of England and the allies.

Near the end of World War II, Germany was losing. Allied troops were closing in on Germany and Nazi-occupied territories. Adolf Hitler thought the perfect weapon could turn things around, and he put his wunder-engineer, Wernher von Braun, on the job of creating a rocket that would send the Americans and allies packing.

Having studied rocketry since his youth, von Braun was on the cutting edge of the science. The treaty of Versailles, signed after WWI, forbade Germany from building artillery, but didn't mention rockets because rockets weren't on anyone's radar as useful weapons back then. So beginning in 1929 amateur German space enthusiasts started designing and building rockets inspired by the publication of physicist Hermann Oberth's book *Ways to Space Travel.* Von Braun took it to the next level and embarked on a secret PhD project with government money. Some say von Braun was all about getting a rocket into space and not about the weaponry, but there was no space program to fund him.

Like Robert Goddard, von Braun knew that using liquid fuel was a more efficient, controllable way to get a rocket to travel great distances. The liquid burns at a higher "specific impulse," which means it gets hotter and gives off more power. Liquids are also lighter than solids, adding a lot less weight to get off the ground. The mechanics in a liquid fuel motor work like this: The fuel is stored in one tank. It's released into the rocket engine combustion chamber with the liquid oxidizer from another tank, where it burns, expands, and turns to vapor. An igniter sparks the mixture to burn. Combustion happens faster than with solid fuel, because there is a lot more surface area exposed. (Think trying to light a whole log on fire with a match versus a pile of sawdust.) It can also be stopped and started again (throttled) by cutting the release of the liquid propellants into the combustion chamber, just like your car's gas pedal.

Hitler wanted to bomb his enemies remotely without worrying about his planes getting shot down and without his enemies having any warning. Because the rocket would travel faster than the speed of sound with the shock cone moving upwards, the victims would have no idea they were doomed until it was too late. Hitler asked von Braun to deliver a rocket that could travel 220 miles, far enough to hit their European foes. The rocket had to carry a one-ton bomb in its payload bay (the rocket's nose), which increased the rocket's weight. To get it to go that far, von Braun designed a rocket that stood forty-six-feet tall, a giant for its time.

Germany was stretched thin by 1943, and the army decided that war prisoners would build the million-square-foot underground rocket factory. The rocket engineering team also faced a labor shortage. After seeing slave labor in an aircraft manufacturing plant, the chief rocket production engineer, Arthur Rudolph, recommended using prisoners to work on the assembly line. Wernher von Braun, who was the project's technical director, chaired a meeting in August where the recommendation was made to produce missiles with the endless supply of prisoners in concentration camps.

The Schutzstaffel (SS—Hitler's elite guard) gathered mostly non-Jewish French, Polish, Russian, and Yugoslav war prisoners and a slew of other nationalities from concentration camps and delivered them to Mittelberg-Dora carved deep into the Kohnstein Mountains. (In mid-1944, Jews from Hungary were brought to the camp, and many Jewish prisoners who were very ill were shipped to Dora in early 1945.)

Freezing cold, without shoes or decent clothing, and almost no food or water, many of the prisoners were forced to live underground for months at a time to dig the tunnels and bunkers. The lives of the prisoners meant nothing because there were always more behind them. Men keeled over and died from exhaustion and starvation. With no other choice, fellow prisoners poured concrete right on top of their bodies.

Yves Béon, an eighteen-year-old French resistance fighter arrived at Dora in 1944 after several months at Buchenwald concentration camp. "Compared to Dora," Béon said later, "Buchenwald was a first-class place. Dora was hell. Dora cannot be compared to any other camp. Even the people who were there were like zombies. It was another world. It was another planet."

Béon and other prisoners were forced to dig a tunnel in the center of the mountain. "When I arrived, the work was [a] horrific job. We stayed months and months inside. We slept there, we ate there." The prisoners were fed a mostly water "soup," "and one small piece of bread and margarine and nothing else." Béon watched friends starve to death within a week of arriving and had no idea how he continued living.

Once the factory was ready, von Braun's brilliant and complicated design was put into production. Whether von Braun completely subscribed to the Third Reich's dogma or not, he *was* a member of the Nazi party, and later, part of Hitler's Schutzstaffel. He soon found himself in a tug-of-war between the army and the SS, both groups vying for power and demanding the rocket scientist's loyalty.

An SS camp commander named Otto Förschner took over the Dora factory in September of 1943. Förschner knew little about managing such an operation, but a lot about forced slave labor and prisoner brutality. There were fifteen thousand prisoners in the camp. Von Braun had no day-to-day interaction with factory operations, but when he did visit, he said, he was shocked. "The working conditions there were absolutely horrible. I saw [Dora] several times, once while the prisoners were blasting tunnels in a hellish environment, clearly worse than in my mind."

The rocket production line was part slave labor and part paid German workers. "There were so many different operations and hundreds of screws," remembers Anatole Gorko, a Polish Jew forced to leave his family behind in Auschwitz in 1944. Gorko lied that he was a mechanic to get out of digging tunnels and work on the assembly line instead. The missiles were so complicated that each one took seventeen thousand parts to assemble.

Gorko recalled his instructions years later: Make the hole, drill the hole half the size of this screw. "Then they put us into operation of the missiles, the V2 bombs, which we didn't know officially what they are."

Something that the Third Reich didn't factor in: Forcing prisoners to build a weapon to kill the people trying to save them *may* cause production issues.

"I started to work, and then came the ideas to me; *It's time to do something*," Gorko said. "We didn't have any arms, we didn't have resistance. We didn't have any possibility to be against the Germans."

And yet, Gorko's brain thought of a way. "I made larger holes for the screw, figuring that they won't last, they wouldn't hold the power. I shortened the screws to a half. And now if you know what *chutzpah*—a Yiddish *chutzpa* is," Gorko said, "I wasn't satisfied with doing it myself."

Gorko was on good terms with the German coworkers on the production line and could communicate well since he had lived in a German-occupied Jewish ghetto for four years. "I was doing the same things they did, and they did the same things that I did. So having this *chutzpa*, I wanted more. I wanted Germans to do the same sabotage." Gorko got close and whispered to his German colleague, "Don't you have any cramps in your palms and your hands?"

"Actually, yes," his coworker said.

"Don't tell anybody," Gorko whispered. "I'm going to give you an easier way of doing it. If you make the holes bigger, it's easier to screw in the screws."

"*Oh das ist eine gute idee, danke,*" the grateful German replied.

"Don't tell the next guy," Gorko said conspiratorially. "It's only for you, because I like you."

Then Gorko went to the next German, and the next. "I had the whole crew doing the same thing. For about two months, most rockets failed on the launch pad. The Germans got sent away and they interrogated them."

The guards had their eye on Gorko too, and if they found he'd sabotaged the rockets, he'd be killed immediately. When they came to him yelling and accusing him of sabotage, rather than feel terror, his *chutzpah* went into overdrive.

"How can you prove this, that I did it? How about all the mechanics that you sent away from here?"

The guards shook their heads and walked away.

"The result—and the talk—gave me so much pleasure that I didn't care if they shot me or not."

Gorko was not alone in sabotaging the V2. There were so many other parts of the rocket to build. The fuel tanks needed welding. The rocket had to be covered by a metal skin layer. There were hundreds of wires in the guidance system to solder. The teenage French resistance fighter, Yves Béon, told the guards he was a welder, although he'd never held a welding torch in his life. Béon soon discovered a way to touch the metal with his welding torch, just enough to look like the weld was solid from the outside but barely holding on the inside. "I had a chance to be hanged if I was caught," he said, but he didn't care. "I knew I would never get out of Dora, so what's the difference."

The propaganda minster called von Braun's creation *Vergeltungswaffe*, the Vengeance Weapon. Hitler hoped this would be his miracle weapon, but the complicated rocket, built by starving men who had nothing to lose, blew up shortly after launch and took out several German aircraft nearby. When they launched an attack on Paris, the rocket blew up in the air. Many more V2s collapsed and exploded on the launch pad. The missiles had a ten-to-twenty-mile error, which made them close to useless.

On April 11, 1945, American troops arrived at Dora to a horrific scene. The Nazis had moved most of the prisoners to other locations. There were very few people left alive, but thousands of emaciated corpses and body parts were thrown into giant piles. Von Braun and his team disappeared.

Q: SO HOW DID VON BRAUN, A NAZI, BECOME AN AMERICAN HERO?

REMEMBER THE JEWISH AEROSPACE engineer from JPL named Theodore von Kármán and the Chinese rocket scientist working for the Americans named Qian Xuesen? They were sent by the US government to meet with von Braun and his team for a super-clandestine job interview. The Germans had two choices: Russia or America. Von Braun opted for America and started creating a new image for himself. He was not shy about telling the world why America was best when he spoke to reporters after his surrender. It was a moral decision for his team, he said. "We wanted to see the world spared another conflict such as Germany had just been through, and we felt that only by surrendering such a weapon to people who are guided by the Bible could such an assurance to the world be best secured."

In a secret mission orchestrated by the United States government called Operation Paperclip (apparently the worst Nazi engineers got paperclips on their files to mark them for exclusion), we brought one hundred and twenty Nazi engineers to America who disavowed Hitler. We gathered their technical papers and one hundred unexploded V2 rockets we found in the factory. The US army worked quietly to sanitize the scientists' résumés, and after a few years, von Braun and Arthur Rudolph were building military weapons for the US Army Ballistic Missile Agency and a giant rocket for NASA that would soon launch American astronauts to the Moon.

In 1961, when I was just a year old, NASA's Saturn I rocket, the daughter of von Braun and Rudolph's V2, was ready for launch. While the V2 was able to fly to the Kármán line (the edge of space at sixty-two miles), Saturn needed to carry a heavy payload and take men and machines all the way to orbit. To get the million and a half pounds of thrust he needed, von Braun came up with a groundbreaking design. Instead of having one engine, he clustered nine liquid engines together for the first stage, called the "booster."

The Saturn I rocket was 162 feet high, weighed 460 tons, and flew up to 17,800 miles per hour to escape the atmosphere. The booster lifted Saturn I from Cape Canaveral eighty-five miles into the sky and across the Earth to land 214 miles away—a near perfect first run.

Von Braun and Rudolph kept building bigger and better Saturn rockets until the Saturn V was ready to carry the Apollo crew and the Lunar Module to the Moon. When that mission was successful, von Braun became an American hero. To the American public, the über-handsome German rocket scientist was irresistible in his fitted suits, sitting next to Kennedy in the president's convertible, on the cover of *Time* magazine—even starring in a cigarette ad.

Von Braun died in 1977 of pancreatic cancer. He was sixty-five years old. Two years later, the US government began a special investigation of Nazi war criminals living in the US that outed Arthur Rudoph for forcing slave labor on the prisoners at Dora. He refuted the allegations, but rather than fight the charges, he gave up his American citizenship and went back to Germany.

Forty years after the end of the war, with most Americans still in the dark about Dora, five silver-haired French gentlemen readied to watch the launch of the Space Shuttle *Challenger* at Cape Canaveral in Florida. The men had asked for tickets to a launch for quite some time through the US embassy in France. They felt connected to NASA's rockets and the launch of the world's first reusable space vehicle, because they lived with the memory of the Dora concentration camp. "All five of us worked on the V2, which was the beginning of rockets," said seventy-nine-year-old Jean Michel, who was captured as a leader of the French resistance and brought to Dora in 1943.

Seeing the giant Saturn V at Cape Canaveral mounted on its side was an emotional sight for the former prisoners. "It's wonderful, magnificent," said Yves Béon, who was now sixty-one years old and had worked to sabotage the V2s with his bad welding. "So many people died [at Dora], we know it wasn't for nothing. Something happy came out of it."

Béon understood the reason that the US government

brought over von Braun's team. "I will never say a single word against America for using Nazis like von Braun and Rudolph, because they helped save the world from the communists," he told a reporter from the *Philadelphia Inquirer.* But what caused Béon immeasurable grief was how the Americans treated the German scientists who were linked to the deaths of so many thousands.

"The question [was] not *using* the scientists," Béon said, "but giving them medals and plenty of money. Another thing which is more important in my opinion that I've seen on several occasions; books made for the kids. In these books von Braun was shown as a great man. A man who was very important in the story of humanity. But that I can't accept."

"After the war, the Allies took the rocket scientists, and Dora was forgotten," said Michel. "Nobody wanted to talk about it because von Braun had become a hero. You can't have a Dora and make von Braun a hero. For us they were just the same as the SS."

Both Michel and Béon wrote books on their experience, and Michel's book is credited for opening the investigation into Rudolph by the US Department of Justice. "I'm sure that most of the American people have never heard of Dora," Béon said. "I would like them to know that these things happened and might happen again."

I was unaware of this history until I was in my late teens. The way the American government and press told the story, it seemed like these brilliant rocket engineers had simply been working for the wrong side. The Cold War was in full swing. Our political leaders thought it best to ignore the horrible crimes in order to showcase the German rocket scientists as patriotic Americans helping us contain the Soviet threat.

Q9: WHAT ENDED THE SPACE RACE BETWEEN THE US AND THE SOVIET UNION?

I WAS AN UNDERGRADUATE, my nose deep in engineering books, when Ronald Reagan took office in 1981. After warming up in the '70s with the Apollo-Soyuz rendezvous, American and Soviet Space relations went into a deep freeze with Reagan's administration. I had always paid attention to global politics, but as an engineer, the world was very much black and white to me. In my naive mind, NASA existed to further our space exploration goals—especially working in the technical academic setting of engineering school. In our minds, we were pursuing "truth and beauty." The injection of politics and international relations into NASA's mission seemed perverse.

Reagan was staunchly anticommunist, (I was in agreement with him, especially since I grew up during the Cold War.) and a big-time devotee to his Christian base. "Yes, let us pray for the salvation of all of those who live in that totalitarian darkness—pray they will discover the joy of knowing God," he evangelized to a meeting of religious leaders in 1983. "But until they do, let us be aware that while they preach the supremacy of the State, declare its omnipotence over individual man, and predict its eventual domination of all peoples on the Earth, they are the focus of evil in the modern world." Soviet-American goodwill evaporated. There were zero talks between our two countries for the first time in fourteen years.

America lapped up Reagan's take, especially Hollywood. Audiences sat in dark theaters, eyes wide, hands gripping the armrests as conniving communists spied on us, shot at our fighter jets, pummeled us in the boxing ring. On screen, the USA always prevailed to cheers from patriotic audiences.

As for space exploration, Reagan was at first more interested in creating orbiting weapons to counter nuclear attacks than he was building up NASA. Reagan had the ear of war hawks at conservative Washington, DC, think tanks who showed him that while the North American Aerospace Defense Command (NORAD) could track intercontinental missiles coming in, we had no way to neutralize them. What we needed was a giant defense shield of sorts. This became the Strategic Defense Initiative (SDI), launched in 1983, a $30-billion, decade-long program where mountains of research failed to make the program a reality.

Critics called it a bad idea for Soviet-US relations and dubbed the program "Star Wars" after the 1977 hit movie. Delaware Senator Joe Biden led the offense. In a 1986 speech, Biden said, "Star Wars represents a fundamental assault on the concepts, alliances and arms-control agreements that have buttressed American security for several decades, and the president's continued adherence to it constitutes one of the most reckless and irresponsible acts in the history of modern statecraft." Biden lacked the fundamental technical education to understand SDI, and he was not alone in his opinion. However, what ended SDI was not a debate on political strategy, but the enormity of the technical challenge and skyrocketing costs. It was too early. Our technology could not yet reliably support the vision within a reasonable budget and time. But the fundamental idea is plausible. Israel's Iron Dome missile defense shield is a small example of what can be done.

The Soviet Union, first under Yuri Andropov from 1982 to 1984, and then in 1985 under Mikhail Gorbachev, found Reagan's goal to arm space with defensive weapons highly suspect. Any cooperation in space exploration was put on hold. That was fine with me. The US was number one, and in matters of the extraterrestrial, we would go it alone. I was in graduate school and planning on applying for an astronaut position.

The first Space Shuttle mission launched and landed in 1981. It was the genius invention and brainchild of Max Faget, who busied himself making paper models of the shuttle way back when Apollo astronauts were launching to the Moon in the vehicle he designed. The Space Shuttle *Columbia* was the first spacecraft that could fly into orbit, reenter the atmosphere, and land in one piece to be refurbished and used again. We one-upped the Soviets on this, but the Soviets were ahead of us in their cosmonaut-tended *Salyut* and *Almaz* space stations. They also performed crew exchanges in space on Salyut 6 and Salyut 7, allowing for continuous occupation over several missions. And Soviet engineers were working on the new and improved *Mir* station.

In 1984, Reagan came around to the idea of space programs that would go way beyond what the Soviets had going on. He tasked NASA with building our own space station with European, Japanese, and Canadian partners. It was a bold idea and a big-ticket item, but Reagan convinced Congress and conservatives to back the plan by promising that the new space station, dubbed *Freedom*, would have a big return on investment. "Just as the oceans opened up a new world for clipper ships and Yankee traders, space holds enormous potential for commerce today," he said in his state of the union address that year. The Republican side of the chamber erupted in applause.

Q: SO . . . WHEN DID THINGS UNRAVEL ON THE SOVIET SIDE?

BY THE LATE 1980s, communism was not going well for its practitioners in Eastern Europe. I was in my late 20s when strongmen holding massive populations hostage under communism began to fall. I loved to see it. I didn't hate communists only because my parents' country had been overrun by Mao Zedong who went to war against educated Chinese citizens. I hated communists and the Soviets because I was a patriotic American, and they were the biggest threat to my country. From the time I was a kid, when my parents overheard me disparaging anyone, I got checked. "We're gonna beat those terrible communists," I'd say to friends as we played war games with our toy fighter jets.

Mom and Dad didn't share my sentiment. "*Xiao Zhong*," Dad would scold me, "you don't know those people. You can't judge individual people by their governments."

Both my parents felt Mao was deeply awful for China and

the Chinese people but did not hold it against Chinese citizens. Mao's totalitarian regime starved and imprisoned millions of people. He and his policies killed more than Hitler's and Stalin's. Mom and Dad had relatives back in mainland China. For most, Mao was a monster. For others, he was a savior delivering them from a brutal military dictatorship and influence from the West. When I went to visit China as an adult in 2006, my mom's cousin in Qingdao had a copy of a famous photo of Mao on his kitchen wall.

In 1989, just two years after I finished my doctorate in chemical engineering, Eastern and Western Berliners tore down the wall that divided free Germany from communist Germany for nearly three decades. Citizens in Poland and Romania rose up and jettisoned their leaders, the latter by firing squad. In June of 1991, the entire "federation" of communist Yugoslavia split into six new (or back to the old) countries, along religious and cultural lines. Can you truly integrate the different religious and cultural societies of a country except by force? I don't think so. After the force is gone, the old ways always seem to spring back up.

By the fall of 1991, I was into my second year of astronaut training. In those days, few of us in the West even knew the Soviet Union was a patchwork of fifteen states with different cultures, religions, and languages forced together by Vladimir Lenin back in 1922. Now the USSR was on the verge of collapse.

At Christmas, I came home from Houston to spend the holiday with my family. I caught up with my sisters and parents while we folded *jiǎozǐ*, and Mom made her delicious salted chicken. They asked me how it was going, and I filled them in on the ins and outs of NASA.

"I was naive. I really thought that NASA, and especially the astronaut office, would be like a family . . . supportive, working together," I said.

I filled them in on the astronaut office, Johnson Space Center (JSC), and NASA brass, all vying for power and position against the backdrop of US and world affairs. Every decision seemed calculated. It could be exhausting and, to me, so unnecessary in the pursuit of what should be an unencumbered exploration into space.

"It's eye-opening for me," I told my family. "It's just like any other human endeavor, I guess." As a young engineer, I was idealistic and naive, and felt that important decisions should be based on technical merit, and that exploration should be for its own sake.

While we filled each other in on our lives, Americans across the country received a surprise geopolitical gift on a scale we couldn't have imagined. Our Cold War nemesis—the terror of our childhood nightmares and the cliché for Hollywood blockbuster bad guys—was no more.

In response to communist countries imploding over the past couple of years, President Mikhail Gorbachev had gifted a bit of capitalism and some personal freedoms for his downtrodden population. But there were few guidelines. Without structure, everything unraveled. He lost control, and his own party ousted him.

Gorbachev was not a flamboyant leader. In his televised final address as president of the USSR, he didn't seem angry or overly emotional. Peering out of heavy-lidded eyes behind wire-framed glasses, his suit rumpled and shoulders uneven, his signature birthmark prominent on his bald head, he made it clear he was done.

"I have firmly stood for independence, self-rule of nations, for the sovereignty of the republics," he said, "but at the same time for preservation of the union state, the unity of the country." He had been in office less than six years. "There is plenty of everything: land, oil and gas, other natural riches, and God gave us lots of intelligence and talent, yet we lived much worse than developed countries and keep falling behind them more and more." Gorbachev blamed the arms race—the decades of military one-upping across the Atlantic as to who could be the bigger bully.

The news startled everyone including President George H. W. Bush, who got just a few hour's heads-up when Gorbachev called him before going live. Gorbachev later said the two talked about future collaborations and agreed to destroy their nuclear arsenals. Bush rushed to get out a response on Christmas night as Americans cleaned up crumpled gift wrap and stacked leftovers in the fridge.

"President Gorbachev is responsible for one of the most important developments of this century, the revolutionary transformation of a totalitarian dictatorship and the liberation of his people from its smothering embrace," Bush said dramatically from the Oval Office. "His personal commitment to democratic and economic reform through perestroika and glasnost, a commitment which demanded the highest degree of political and personal ingenuity and courage, permitted the peoples of Russia and other Republics to cast aside decades of dark oppression and put in place the foundations of freedom."

To my Chinese American family, the mammoth geopolitical shake-up was purely interesting. For me, I liked seeing communism fail, but I didn't see it affecting me or my job with NASA. Talk in the hallways of NASA was of disbelief and wonder of what lay ahead. Pretty much all the military folks were happy. To a large degree, I agreed. However, I was an astronaut now. There was a very cool history of collaboration with the only other "game in town" when it came to space exploration. I was curious of what this might mean for working together in the future.

Q: WHAT HAPPENED TO THE SOVIET SPACE PROGRAM?

THE COSMONAUTS WERE SCREWED. We got news that the workers in Moscow's Mission Control and the Baikonur launch center were on strike because they weren't getting paid. Engineers, doctors, and other specialists were driving their personal cars as taxis to make ends meet. But there was one story in particular that was beyond wild.

Earlier that year, in May of 1991, cosmonaut Sergei Krikalev launched on his second mission to *Mir.* The commander, Anatoly Artsebarsky, was on his first mission, and they were joined by a twenty-seven-year-old female chemist named Helen Sharman who was the first British person to go to space. Sharman stayed for a week conducting experiments. Six months later, Artsebarsky left and was replaced by Alexandr Volkov. A third cosmonaut from Kazakhstan arrived as well, but it turned out he wasn't trained for a long-duration mission. Krikalev, already five months in, volunteered to stay another six months. He had a baby daughter at home, his first child with his wife, Elena, who worked as an engineer at mission control. But there really wasn't a choice. He couldn't leave Volkov alone.

Krikalev was just two years older than me and on his second mission. Volkov was a veteran on his fourth mission. Both loved being in space. Krikalev especially loved the *Mir*, a ragtag spacecraft known for loud rattles and stifling humidity. The athletic aerobatic pilot found the medley of sweat and mold in the air reassuring. Being in space for six months was normal protocol.

"The long-duration stays at *Mir* had the objective of better knowledge of extended missions in space," Volkov said in an interview. "We also tested recycling systems. We are developing a system of extracting oxygen from urine." But for Krikalev, who was now doing double time, he thought about his high school track coaches who told him to pace himself for the duration of the race. "When you realize that your distance is much longer, you have to change [your] mindset and actually pace for your activity and especially to be able to finish [the] longer distance," he said, "And it was really a change in mindset."

Mir could theoretically fit twelve cosmonauts, but it was crowded with equipment and air ducts and various tubes and wires that emerged from the cabin walls and cluttered every surface like a hoarder hovel. There were crew quarters, lab spaces, and a small gym area with an exercise bicycle and rubber bands for resistance exercises.

The cosmonauts spent time staring out the window and photographing the Earth, fixing broken things inside and outside during spacewalks, and conducting experiments. It wasn't glamorous work. "We never considered ourselves as 'members of an elite,'" Volkov said. "The cosmonauts always considered themselves as the 'workhorses of space.'" Down below, however, the government and the public treated cosmonauts like celebrities. Cosmonauts and their families lived in nice government-given apartments. Senior cosmonauts lived in huge homes on "cosmonaut row" just outside of Star City, and they were gifted private cars—a far different lifestyle than the average Soviet citizen.

Things were smooth sailing on *Mir* until December 26, 1991, when Krikalev and Volkov got some distressing news. From a *glavny* (the person in charge of communicating with the cosmonauts, what we call CapCom at NASA) in mission control, they learned that they may no longer be employed. There may be no more cosmonaut corps. Worst of all, and almost incomprehensible—they no longer had a *country*.

Normally, Krikalev was cool as ice. Now, his heart and mind raced.

How are my wife and my baby daughter? he worried.

From Western news, they heard that the Soviet Union had broken up, but they had few details. Weeks went by.

For my astronaut class at NASA, it was wild to think about. We learned that people at the Russian launch center broke into the food because everybody was hungry. No one was getting paid. At one point, they threw empty food containers in the supply ship and launched the garbage to the space station.

In the coming weeks, their colleagues at mission control, unpaid and without a Soviet boss to direct them, shut the lights off and abandoned their posts. The last two Soviet citizens circling the globe were merely an afterthought in the chaos of an imploding global superpower.

Still, Krikalev was more worried about everyone on the ground than himself. "They say it's tough for me—not really good for my health. But now the country is in such difficulty, the chance to save money must be the top priority." He was also glad he had gained a lot of prior experience, ". . . sooner or later you face some problem, and you realize that with some experience you can troubleshoot the problems during flight, and you can do a lot of things yourself," he said. "Because in the beginning you rely more on your partner, you rely on the ground. But [with] your experience, you became more and more self-sustainable. So you're able to face the problem and solve the problem even if nobody can help you."

The Russian team was trying desperately to make ends meet to support a skeleton crew on the ground and to keep Krikalev and Volkov alive. They had already sold seats on the Soyuz to astronauts from other countries. Some said that they were offering to sell the *Mir* space station to NASA. (We didn't want it.) There was almost no money to launch a supply ship full of food to the stranded cosmonauts. Even if they could find the money, the Russian economy was in a free fall, and many common goods were nowhere to be found. Krikalev dreamed of lemons and honey. Still, he felt grateful. "I was doing my job and was more worried about those on the ground—our families and friends—we had everything we needed." Physically, he knew he would survive. Mentally, it was an enormous challenge. "Do I have enough strength? Will I be able to readjust for this longer stay to complete the program? Naturally, at one point, I had my doubts."

Krikalev had reason to be worried. On the ground, the entire space program was in shambles. That meant Krikalev's wife, who tried to communicate with her husband from mission control as often as she could, was also struggling. "I tried never to talk about unpleasant things because it must have been hard for him," Elena said. She was preoccupied by the fact that her husband and his crewmate may not even be permitted to land at the nominal site near Arkalyk, in what was now

Kazakhstan—an independent country beefing with the new Russian Federation.

Things finally got settled, and on March 25th, 1992, both men arrived on Earth in one piece. Their USSR passports were no longer valid. Krikalev was now a citizen of Russia. Even Krikalev's hometown had changed its name from Leningrad to St. Petersburg. Volkov was now a citizen of a completely different country, Ukraine.

And just like that, the last two chess pieces in a thirty-four-year-long space race between the Soviet Union and the United States of America were out of the game. The few remaining groundcrew carried Krikalev and Volkov out of the capsule, as they were unable to walk on their own because their balance was off-kilter. After 311 days in space, Krikalev was especially deconditioned.

It didn't matter. Krikalev was happy and relieved to be home. He took the ongoing turmoil of his new country and its human inhabitants in stride. "I think that everyone who has been to space gains a global perspective, you stop seeing things simply as national issues," he said. "I know that people sometimes start to argue with no good reason; but especially when you are in a harsh environment, you rely on each other and try to help each other. That is how we live in space, and that probably can be a good example for people on the ground."

My NASA colleagues and I half-heartedly applauded Krikalev and Volkov and the new freedoms of Russian cosmonauts and aerospace engineers. Some we knew by name, but we didn't necessarily like or trust them. We were glad we weren't in the same boat. In comparison, we had a fat budget. The members of my astronaut class were wide-eyed and excited about our futures. We focused on the Space

Shuttle since Reagan's Space Station *Freedom* was already having big programmatic issues and was likely to never get off the ground.

We were a mix of confident and motivated men and women who were either military pilots or rising engineers. None of us were well-versed in political thought or global affairs. Still idealistic, I thought that the space program was the space program. How dare anyone play politics with the space program?

It may have done me some good to look a little deeper into politics, because a new president took office in 1993. Bill Clinton realized that it wasn't only hardware that the newly formed Roscosmos was auctioning off to the highest bidder. The intellectual know-how of Russian cosmonauts, rocket scientists, nuclear physicists, and space engineers were for sale as well and looking for a benefactor.

After a dedicated childhood of commie hating and working my butt off to be an astronaut, politics played me after all. In 1994, I thought my eyes were deceiving me when Sergei Krikalev, the highly-experienced and famous cosmonaut, and Vladimir Titov, another veteran, walked onto the campus of Johnson Space Center with their Russian accents to report for duty.

What the actual hell?

PART THREE

MAIN COURSE

The Meat and Potatoes of an Astronaut's Life

Q10: HOW DID YOU GET HIRED TO BE AN ASTRONAUT?

IN 1986, I WAS a graduate student at the University of California, Santa Barbara, working on my PhD in chemical engineering. I had a feeling that it was premature to apply for the astronaut program, since I hadn't graduated yet, but I couldn't help myself. I wrote a letter to the Johnson Space Center and asked for an application. A manila envelope arrived, thick, heavy, and smelling like opportunity.

NASA was having a moment—in a good way. In 1984, to get people hyped up about big-ticket items at NASA, the Reagan administration and the agency got together and brainstormed a big PR idea: *Why don't we send a schoolteacher to space? Someone relatable who can explain in a digestible way all the complexities of going into orbit.*

A national competition with an eleven-page application found the perfect candidate; Christa McAuliffe, a thirty-six-year-old social studies teacher from New Hampshire. With a thick Yankee accent, a toothy smile, and mountains of enthusiasm, she was a big hit. In a *Today Show* interview in July 1985, the mom of two told host Bryant Gumbel that getting the position out of 11,400 candidates was like winning the lottery. "I still can't believe that I'm gonna actually be going into that shuttle. It just really doesn't seem possible. Maybe when I'm on the launch pad it will."

McAuliffe had been a space fan her whole life, and she wanted to bring the wonder and excitement of space travel back to students. Was she worried? Gumbel asked.

Not at all.

"Spaceflight today really seems safe." McAuliffe was more concerned about getting class assignments written for her substitute and making sure her husband and kids would be okay while she was gone.

The country loved Christa. There was a small subset of Americans who were grumpy that one of the seven most coveted seats on planet Earth would be given to someone who would train for a mere 114 hours to ride in it, compared to the full year that an astronaut would normally train for before going into space. If you guessed that I was one of those grouches, you're not wrong. But I knew the long game for Reagan and NASA was the Space Station *Freedom*, and that would mean a whole bunch of new missions for astronauts who had worked tirelessly for their opportunity to go to space. For that to come to fruition, Americans needed to care about space.

I was filling out the astronaut application in my student office on January 28, 1986. It was a process. Everything was done via snail mail: transcripts, letters of recommendation, medical reports. You had to be in excellent health, which I was. I was just twenty-five years old, but my engineering PhD would qualify me (once I had it, and I was almost there).

I didn't tell anyone I applied to NASA. I'd made that mistake back in undergrad. I was at the frat house chatting with a fraternity brother about what we wanted to do after college.

"I'd like to be an astronaut," I told him.

He erupted in laughter. Then he went and told a bunch of our Phi Tau brothers.

"Hey, Leroy, I want to be a fireman," one guy said as he passed me in the hall, cracking himself up.

Steve Valles stopped by my room and stuck his head in. "I want to be a policeman!" he yelled.

They ribbed me for a few weeks, but it was mostly good natured. (I got the last laugh, obviously, and Steve and I laugh about it to this day.) I didn't think it was that much of a reach. If I'd wanted to play in the NBA, I knew that wasn't going to work out. I'm not very tall or particularly athletic. But I knew I was smart enough and in good enough health that I could at least *apply* to be an astronaut. I was penning my cover letter about how I thought I would be a good fit as a mission specialist on the Space Shuttle and future Space Station *Freedom* missions when a colleague ran into my office.

"Leroy, did you hear?"

"What?"

"The Space Shuttle! It just blew up!"

I turned the small TV in my office to CNN. A few friends came in, and we watched in disbelief. Twenty-four-hour news was a new thing, and CNN filled the time by repeating the explosion again and again and again—a giant orange fireball, followed by cascading white clouds of combustion and debris falling back to Earth. Christa McAuliffe and the six others on board had flown just seventy-three seconds of their mission.

In the back of my head, I heard my mom. *Too dangerous, Érzi. You need to be an engineer. It's much safer.* I couldn't deny she had a point, and here was proof.

It came out later, that on the morning of the disaster, one of the contractors recommended scrubbing the launch. Cape Canaveral, Florida, was uncharacteristically *freezing.* The temperature was in the high twenties to low thirties, colder than any other launch to date, which the engineers argued could affect the solid rocket boosters. But NASA was under pressure from the White House because Reagan was planning on touting the first schoolteacher in space during his State of the Union address the following week. On the launch pad, the low temperature made the O-rings (rubberlike seals between the solid fuel booster rocket segments) too brittle to make a tight seal. On takeoff, the burning fuel escaped and burned into the liquid fuel tank like a blowtorch. Seventy-three seconds after launch, the entire Shuttle system exploded forty-six thousand feet above the Atlantic Ocean.

It was tragic, but it didn't deter me from wanting to go to space. Human error, hubris, politics, and bad decisions will always play a part in space exploration. Also, NASA had a good track record of fixing things and moving forward. There hadn't been a fatal mission since the Apollo 1 fire killed the crew during a routine test, nearly twenty years earlier.

A few weeks after the disaster, I put a stamp on the envelope with my finished astronaut application, walked to the post office, and handed it to the clerk. And, in the age-old tradition of first generation Asian American kids, I told my family nothing about it.

A few months later, I got a postcard back—the '80s equivalent of an automatic rejection email. "You don't quite meet the qualifications yet," it said, "so you weren't considered. Finish your degree, get some more experience, and you're welcome to reapply." It was fair, and I wasn't surprised. I took the fact that they asked me to reapply as a positive.

For my PhD, I was working on bread-and-butter chemical engineering—the dynamic behavior of chemical reactors. I enjoyed the research, but I knew I didn't want to work for a chemical manufacturing company. I was steering more toward academia, maybe becoming a university professor. But my eye was still on the astronaut prize, even though it was a long shot. NASA's requirements at the time were a bachelor's degree in science or engineering. I knew that having a masters' and a PhD would help my application. NASA's philosophy was that if you're technically good, if you've shown that you can learn in an academic setting and beyond, that you know how to use the scientific method and think logically, then you can learn how the Space Shuttle works. You can learn how the engines work. You're smart enough to figure that out.

In order to get some industry experience to bolster my chances in the future, after I finished grad school in 1987, I took a job with a company called Hexcel Corporation. The company specialized in creating honeycomb material and composite materials for the aerospace industry—superstrong, lightweight carbon fiber composites. A year and a half into my job at Hexcel as a Research Specialist, I wrote to NASA for another application to the astronaut program. Times had changed for NASA after the *Challenger* explosion, and the fate of Space Station *Freedom* was a big question mark under the new George H. W. Bush administration.

Fortunately for me, becoming an astronaut in the late '80s was far different than in the 1960s and '70s. If I'd started two decades earlier, my engineering degree and job experience wouldn't have meant squat in the astronaut ranks. They were all military pilots with impressive flight logs. And since Mom put the kibosh on my test pilot career when I was six, I would have never stood a chance.

What changed?

The Space Shuttle.

If Americans were going to be in space, we wanted to make it worth our while. Since the Space Shuttle was the only certainty, it was an easier sell to tout all the important science and discoveries that could be made on board. NASA wanted to be on the cutting edge of all kinds of things, but especially *material science.* Their dream was to create novel materials by conducting experiments on the Space Shuttle—like a factory in space cranking out wild new stuff that was difficult or impossible to make on Earth. We hoped to form metal alloys that would have different properties because they didn't have to form under the pull of gravity. Some researchers hoped the microgravity environment could allow protein crystals to grow larger and purer in space (free of the defects that gravity would create) to help develop new medications.

With a spacious cabin that held up to seven humans for up to two weeks (with only two of them needing to be test pilots), there was now a big need for scientists and engineers in the astronaut ranks, especially material scientists. This is how astronauts became divided into two categories: pilots and mission specialists. *Note:* The two pilots on each mission are called the commander (who is the lead in the left seat) and the pilot (who's in the right seat) because no one wanted to be called a copilot. And that goes all the way back to the Gemini program the first time there was more than one astronaut onboard.

At the time, the chief and the deputy chief astronaut at NASA were from military backgrounds. Everybody reviewing my application where I listed my current employment as a materials process engineer said, "He's a materials guy." My application got put in the material science pile, even though what I was working on at Hexcel had nothing to do with material science research. In fact, I'd never taken a material science class in my life. Material scientists looked at things at a molecular level, an atomic level, and invented ways to stick things together and form new substances. I was the engineer to those creations. I took what they made and helped figure out how to create useful things with the materials. I knew it wasn't a problem. I knew I could learn whatever I needed to know.

Months went by after I sent my application. Finally, I got a call early on a September morning.

"Leroy, this is Duane Ross, I'm the head of the selection office for the astronaut program. We'd like to fly you to Houston and interview to be an astronaut."

He'd woke me up out of a dead sleep. It was 6:00 a.m. California time.

"Uh," I said. "When do you want me to come?"

"Next week."

Maybe it was cobwebs in my head or my subconscious not knowing that this was the most important phone call of my life to date, but without thinking, I said, "Well, I have tickets to the Reno Air Races next week. Uh, could we do it the following week?"

There was a pause on the line.

"Uh, yeah, we can reschedule it," Duane said.

And that second, my conscious mind woke up. *Leroy! What the fuck did you just say?? Get yourself together, man!*

During the NASA astronaut recruitment process, twenty prospective candidates come to interview at the same time. Over the course of several months, they interview one hundred total. Thousands apply. Many years they don't even open the Astronaut Candidate Training program. *So why did I ask to reschedule a once-in-a-lifetime opportunity? WHAT IS WRONG WITH ME???*

"You know what? I can make it. It's fine."

Duane said, "No, it's fine. We can defer it."

The word *defer* reverberated in my head and gave me so much dread.

"No, no, no. I can make it. For sure. I can. I can totally make it."

Q: WHAT WAS THE ASTRONAUT INTERVIEW LIKE?

I FLEW TO HOUSTON the next week. It was September. I emerged from the Houston Hobby airport into a wall of stifling humidity. My shirt stuck to my skin; it was almost hard to breathe. My California-calibrated system was in shock. *How can people live here?*

Johnson Space Center, NASA's ground zero for everything astronaut, arrived in this southern inferno just a year after I was born. *But why here?*

With JFK rearing to go to the Moon in the early 1960s,

he needed someone to run NASA who could handle all the moving parts that such an extraordinary mission would demand. James Webb was fifty-four years old, semiretired, and involved in a whirlwind of projects and boards when he was asked to head NASA in 1961.

One of Webb's first orders of business was to find a location where all the activity around manned spaceflight would live and breathe. With tons of money and jobs coming along with a big NASA installation, several cities vied for the opportunity, including St. Louis, Missouri, Boston, Massachusetts, and Lawrence, Kansas.

"Houston had a very large technical capability," Webb said. There were aerospace contractors already there, like McDonnell Aircraft Corporation, and lots of brainpower coming out of Rice University and Texas A&M. Webb also wanted the site to have access to water so they could move giant things like rockets around more easily. And they needed land. It just so happened that Rice University had a big ranch that they figured could get them a lot more clout as a space facility than a cow pasture and leased it to NASA for a rumored one dollar a year.

Heat and humidity be dammed, Houston it would be, Webb decided. "The basic reason was to get the space program done in the most effective way possible and to build scientific engineering and managerial strength for a long time into the future. We were not looking at the cheapest way of making stuff in tents or building poured concrete buildings," he said. "We realize that when man left the air, when he had an engine big enough to leave the air and the Earth and to move around in space, you are entering a new and unlimited arena, and that we were not just constructing the fastest, quickest way to get a few payloads into orbit. We were building permanently."

He was a thoughtful visionary. Thank you, James Webb.

I arrived at a dingy commuter hotel NASA had booked me on a Saturday in 1989. As soon as I opened the door, I watched a cockroach run across the pillow on my bed.

Okay . . . This is awesome.

The next morning at 8:00 a.m., everyone in the interview group gathered in a conference room. There were twenty of us. It was the late '80s, and the group was still dominated by white male military pilots. I was the only Asian there.

One guy in my interview group was Rich Clifford. He was close to forty with prematurely white hair but still young-looking. Rich was already working for NASA (on loan from the Army) as an engineer on different parts of the Space Shuttle. His title was Space Shuttle Vehicle Integration Test Team engineer. He and I hit it off the day we first met, and he later became my best friend in our astronaut class.

We spent the day taking standardized tests, an IQ test, a psychological battery, cognitive tests on logic and memory—you name it. It took hours and hours. At the end of the day, we had a little social with the members of the committee who came out to meet us at the hotel. One of them was John Young, one of, if not THE greatest astronaut of all time.

John Young had the look of an English professor, and even at sixty years old with his hair white, he had a youthful appearance. Young walked on the Moon. He flew before that with Gus Grissom on the first Gemini. And he was the very first Shuttle commander in 1981. Young was a hero for all astronauts and anyone who knew anything about astronauts. He was a cool, collected, navy fighter pilot (they call themselves Naval Aviators) and the most impressive guy I'd ever met.

I was acting low-key in his presence, but I couldn't have been more in awe if I was Catholic and he was the Pope. Partway through the conversation, a guy from Los Alamos interrupted John Young and said, "What was your name again?"

"John Young," he replied.

And the guy said TO JOHN YOUNG, "And . . . *what do you do?*"

Me and an Air Force guy froze.

Oh my God, you don't know who John Young is?

Needless to say, that guy was not selected to be an astronaut.

The rest of the week seemed like one endless medical exam, the most thorough medical exam you can imagine. Men and women in white coats drew blood, scanned, x-rayed, checked our psyche, our balance, our urine. They did things I never had done to me before.

"We always like astronaut interviews because you guys will do everything we want you to do with smiles on your faces," laughed the Chief Flight Surgeon. We thought that he might have been a bit of a sadist after the countless tests and poking and prodding, but he actually turned out to be one of the nicest people I've ever known.

We were also encouraged to visit the building where the astronauts' offices were located. I approached their floor, hoping to catch some of the astronauts on the selection committee.

It was a little intimidating. Up until then, there had been only seventy-eight astronauts who had gone to space. It felt like hallowed ground on that floor. While I walked around, hoping to randomly bump into one of my idols, I checked out all the names on the office doors.

There was not one Asian name at the time. A poster on the wall displayed every NASA astronaut who ever flew.

I scanned it quickly. There had only ever been one Asian American to fly in space: Ellison Onizuka, a military flight test engineer born in Hawaii to Japanese parents. Tragically, he died in the *Challenger* explosion. It was his second mission.

Rather than let the lack of Asian representation intimidate me, I looked at it practically. I certainly wasn't counting on my Asian-ness giving me any kind of advantage (it had been a huge disadvantage while applying for engineering school), and I knew that NASA was focused on a meritocracy at that time rather than fulfilling certain quotas.

The actual astronaut interview was only one hour long. I had mine near the end of the interview week.

When it was my turn to be interviewed, the head of the selection board said, "By the way, you get a gold star for keeping your essay to one page."

"Really?" I said. "People write more?"

"Oh, God. We've had people that write thirty, fifty, one hundred pages. Nobody wants to read that!"

"You're welcome." I smiled.

It was a twelve-member board consisting of senior astronauts, some in management from both Houston and other centers, and some from the NASA headquarters. There was just one woman who was on the committee, Carolyn Huntoon. In 1989, while the recruitment of women at NASA was up, there were few women in senior positions.

Carolyn had been with NASA from the very beginning of manned spaceflight. She started full-time in 1968 when she helped set up a medical laboratory at Johnson Space Center and oversaw thousands of hours of groundbreaking research on how spaceflight affects the human body. At the time of my interview, she was the Director of Life Sciences at the Johnson Space Center, and she later went on to be the first female NASA Director at the center.

The conference table was set up in a T arrangement. I was in one of the corners of the T so that I had my back to a couple of members of the committee. I think they did that to try to rattle me.

What do you think of working here? Would you do this here? Would you do that here?

They were easy questions for the most part. Then they threw me a curveball:

"If you had all the resources in the world, what would you do to overhaul the education system in the US?"

I answered the best I could.

"I would put money into technology and computers. That's obviously the future," I said. "I would also try to bring some disadvantaged demographics and regions along, starting at a young age. I think that is the key."

I wasn't nervous. There wasn't time for everyone on the board to ask a question. It was interesting that, though I was in Houston for an entire week, engaging in countless tests and health checks and social events, the interview portion was just one hour long. That was it.

At the end of the week, we had a social night at a Cajun barbecue place that was a hangout for the folks from JSC. It was a hole in the wall next to the Ellington Field airfield that NASA uses. I think this was where the real selections were made. They wanted to see how we would interact with them. We ate Cajun barbecue, red beans and rice, boudin sausage, and brisket. I had a couple of beers. I talked to a bunch of people and met some astronauts that I knew by reputation and others who were on the board. And then, I went back to California without a clue as to how well I did.

Q: HOW DID YOU FIND OUT YOU GOT IN?

JUST A FEW DAYS after I got home, I got a call from the OPM (United States Office of Personnel Management) and the guy said, "Wow, they must be really hot on you because you just interviewed, and you're being put in for a security clearance already."

I took that as a good sign. Then I heard nothing. Months went by.

Finally in January of 1990 I got a phone call from Carolyn Huntoon.

I'd learned my lesson to shake myself awake for any early morning call.

"Leroy," Carolyn said. "You did great, but I'm sorry to tell you that we didn't select you."

Ouch. "Okay," I said.

"But you're welcome to apply again," she said.

"Thank you, Dr. Huntoon. I'm disappointed, obviously, but I maintain the highest enthusiasm for the program and will definitely reapply."

I was sad. But fortunately, I was enjoying my job at Lawrence Livermore National Lab doing research work on processing thick-section composite materials for aerospace and other applications.

On the 17th of January 1990, my phone rang again. Six a.m. It was the chairman of the selection committee, Don Puddy. Puddy was an engineer and heavy hitter at NASA since the Apollo days. I was immediately wide-awake.

"I understand you had a conversation with Carolyn yesterday," Puddy said.

"Yes, sir," I said. "I'm disappointed, of course. But I maintain the highest enthusiasm and would like to keep applying."

"Well, how'd you like to join this class?" Puddy said.

Zero hesitation.

"Absolutely! I would love to join this class!" I was smiling from ear to ear.

And that was it. Delivering epic news to your Chinese American family and friends looks nothing like those American families on TV that gas you up, throw you a party, and brag to all their friends.

I tepidly called my girlfriend, Janny (also a first-gen Chinese American), who was in California doing a medical residency.

"Congratulations, Leroy. I know that's what you wanted," she said, sounding mildly excited for me. We both knew that this meant I had to move to Houston, but neither of us mentioned it. Things between Janny and I were rocky. We'd started dating when we were sophomores at Berkeley. By this time, we'd been together for seven years, and we were going in separate directions. It seemed to me that Janny had a lot of anxiety, which I didn't understand or know how to deal with. Her parents were both medical doctors, so she felt immense pressure to be one too.

Next, I called Mom and Dad.

"Congratulation, *Érzi*," Mom said. "I'm so proud and excited for you. But this is dangerous, right?"

"Mom, don't worry," I said. "It'll be okay. I promise."

"I guess you're going to leave engineering and go off and do this for a while," Dad said. "But you should minimize how much time you do this. If you could fly in one year and get it over with, you can get back to work without too much loss of continuity in your real job."

"Dad," I said, "I don't think it works that way. The basic training is a year. Then it's going to be a few more years before I get a chance to fly."

Dad *harrumphed* on the other end of the phone.

At twenty-nine years old in 1990, I was NASA's youngest astronaut. I was not the youngest ever, but I was close. Charlie Bolden, an astronaut and member of the selection board, later told me, "You were very impressive, but you were so young that we decided to defer you and have you apply again. But then when this position became available, you were the next one on the list. So we brought you in." Charlie was a Vietnam veteran pilot who became a good friend of mine and went on to become Administrator of NASA under Barack Obama. *So cool.*

Janny broke up with me before I left, which was a huge relief. I came to Houston with a fresh start and began a new life as an astronaut candidate—embarking on my life-long dream. Out of the 1,945 qualified applicants who applied that year, NASA invited 103 candidates for interviews and medical exams. Of that 103, just twenty-three of us were chosen for NASA Astronaut Group 13. It felt great to be a part of that group.

Q11: WHAT IS ASTRONAUT TRAINING LIKE?

I FELT COMPLETELY CONFIDENT about flying to space. I'd been practicing since the day I built my Apollo capsule in Dad's garage. But there was part of the astronaut job that I'd never thought about, and for that part, I didn't feel ready at all. As soon as they announced my class in January 1990, my phone started ringing with interview and speaking requests. I wouldn't be starting officially until July.

I called the astronaut selection office.

"I'm getting all these people calling me for interviews and to give talks," I said, wanting some guidance.

"Well, that's totally up to you, Leroy," said Duane Ross, the Director of the Astronaut Selection Office. "You can do them if you want. But you don't have to, because you don't work for us yet."

I realized that once I was a full-fledged astronaut, I'd be expected to do a lot of public speaking and media interviews. The American people paid my salary and for all the machines that would send me to space. They deserved to know what NASA was doing. My stomach knotted up just thinking about speaking in public or talking to reporters. I'd never been able to fully shake the time I had to give a report in seventh grade. I'd rehearsed my talk in my head but still ended up frozen on the spot. In college, whenever I had to speak about a project to the class, I was always nervous, and during technical conferences in graduate school and in my early career, my public speaking skills improved, but it was not fun.

After several press requests, I said to myself: *You know what? I'm going to do all these interviews and speaking gigs so I can practice and get better.*

I accepted every invitation that came my way.

In a big auditorium, I spoke to students and faculty at North Carolina A&T State University (where Ronald McNair, another astronaut killed on the *Challenger* accident, had earned his bachelor's degree) about a collaboration between the university and NASA's planned mission to Mars. I was given plaques and awards from every local group that had heard of my hiring. I had to "say a few words" after rubber chicken dinners. I said yes to TV and print interviews.

Then I put myself to the ultimate test. A second-grade teacher at Portola Avenue Elementary School asked if I would come talk to her class.

How hard can it be?

I was sure the kids would ask me simple spaceman questions.

"Dr. Chiao?" a little girl asked. "How much money do they pay you?"

Gulp.

I'd never even asked the selection office, because it would hardly have mattered to me. This seven-year-old had more finance acumen than I did.

"You know, I don't know how much I'll get paid," I admitted sheepishly.

"Dr. Chiao," a young boy piped up, "what's your very favorite planet of all time?"

"Earth!" I quickly responded and was relieved by the chorus of *me too*s and a few screams of *Mars! Jupiter is better!* It was a ten-minute presentation that seemed like an hour. The teacher came to save me.

"Say thank you to Dr. Chiao, students."

"Thank you, Astronaut! Good luck going to space!"

I wiped the sweat from my brow and got the heck out of there.

Side note: Talking to kids and students about being an astronaut and going to space has been one of my very favorite things to do for my whole career. I love seeing their wonder and answering their very straightforward questions.

By the time I was packing up to move to Houston, I could get up and speak about almost anything space-related without even a hint of nerves.

Note: Chinese parents don't allow their kids to say, "I'm not good at this or that." You just *get* good at it if it's something you need to do. Goes to show, you can learn lots of things you don't have a natural affinity for if you work hard and keep learning.

Q: WHAT WERE YOUR FIRST FEW WEEKS LIKE?

I WROTE A $3,000 personal down payment check, and I bought a brand-new three-bedroom house in a subdivision near NASA. I was stoked. I reported for work at the Johnson Space Center on July 16, 1990. The first day, my twenty-two new colleagues in astronaut Group 13 and I went through the formalities of being sworn in to be a civil servant of the United States of America.

Raise your right hand. Repeat after me . . .

> "I, Leroy Chiao, do solemnly swear that I will support and defend the Constitution of the United States against all enemies, foreign and domestic; that I will bear true faith and allegiance to the same; that I take this obligation freely, without any mental reservation or purpose of evasion; and that I will well and faithfully discharge the duties of the office on which I am about to enter. So help me God."

The H.R. briefings were long and boring. I got a welcome packet and a NASA mug. I filled out forms for my retirement options and health insurance. Then we got issued our blue flight suits, changed into them, and went to a press conference on a raised stage where the public affairs folks introduced our class to the reporters.

My ethnic background made me somewhat interesting to the press. I was the first full-heritage Chinese American professional astronaut and only the third Asian American. But I wasn't the only *first* in our class. There was the first Latina woman, Ellen Ochoa. Ellen was a mission specialist like me.

Unlike my mom and dad, Ellen's parents didn't have college educations. Ellen, a second generation Mexican American, was brilliant, self-motivated, and rose to the top of her class at San Diego State. She went on to complete a PhD at Stanford in electrical engineering. Ellen was just a couple of years older than I was, but at the time she applied to be an astronaut, she was the Chief of the Intelligence Systems Technology branch and managing a team of thirty-five engineers at NASA's Ames Research Center in California.

The press was also delighted our class had the first woman who would pilot the Space Shuttle. Eileen Collins flew for the Air Force. Back then, women weren't allowed to fly combat aircraft, but they could become test pilots, a prerequisite for all NASA pilot astronauts. When Collins sent in her application, she figured she'd shoot for a mission specialist position. She'd earned two master's degrees while working as an Air Force pilot, one in operations research and another in space systems management. But when she got the call from John Young, Collins asked if it was for mission specialist or a pilot position. Young laughed. "Pilot! You will be the first woman to pilot the Space Shuttle!"

The newspeople asked us questions and snapped photos.

"How do you feel, Leroy?" asked one reporter.

"I still can't believe it's actually happened," I said, completely unfazed by the press, enjoying this bit of newfound celebrity. "It's been kind of my boyhood dream."

My class arrived at JSC at a prosperous point in NASA's history. A few years earlier, things were rough in Reagan's second term with the *Challenger* explosion, waning public interest in space, and the Space Station *Freedom* program on life support and soon to expire completely.

Fortunately for me and my classmates, President George H. W. Bush had big space goals, which made all of us feel like we had lots of exciting missions in our future. In May of 1990, while I was looking for houses in Houston and wrapping up job stuff at Lawerence Livermore, Bush was speaking at graduation ceremonies around the country touting the future of the space program. NASA's budget would go up by 24 percent, a massive increase, he said. The money would pay for what would be the International Space Station, a lunar base, and by 2019, we would send the first astronauts to Mars. "It's time to open up the final frontier," the president said. "There can be no turning back."

"How will you get there?" one reporter asked.

"You have to go fairly fast," Bush joked. "It's a long way out there."

This was my dream. *We are going to Mars!*

It was the number one topic of conversation when my class all came together.

"You think one of us might get to go to Mars?" I asked the group.

"Maybe. And hopefully the Moon before that!" Rich Clifford, another mission specialist said.

Rich was my best friend in our class. We hung out together, ate lunch together, and worked out together. Richard was almost a decade older than me and had a very different career than I did. He was a West Point graduate who flew helicopters and fixed wing aircraft. His Army pilot background technically qualified him to be a shuttle commander. But the Navy and Air Force pilots were in charge, so he came into the ranks as a mission specialist like me. Although our backgrounds were quite different, we became fast friends.

It was a thrilling time. We didn't hide our enthusiasm and excitement. We were NOT *too cool for astronaut school.*

Our class consisted of pilots and mission specialists. Pilot astronauts were going to be flying the Shuttle in the two seats in the front of the spacecraft flight deck. We call these astronauts the "front seaters." Then we had mission specialists. We were the research engineers, scientists, or medical doctors who would be doing things like scientific experiments, spacewalks, and manipulating the robotic arm.

One of the first orders of business was to come up with our class nickname and mascot. A veteran astronaut, Bill Shepherd (astronaut Group 10) told me to pick a name that people wouldn't forget. His '84 class was the Maggots, which was still a legendary nickname. We were the 13th class, the superstitious "unlucky 13." I decided a black cat was the perfect mascot and during one of our first class meetings in the main astronaut office conference room, I volunteered to be in charge of designing our class patch. Not everyone was enthusiastic.

"Then what should our class nickname be?" Ellen asked.

I thought about the cigar-shaped globs that cats hack up.

"The Hairballs!" I said, delighted with the image

"Gross." Ellen recoiled. Several others had the same take.

"That's precisely the point!" I countered. Bill had told me that some would be hard against something gross like maggots or hairballs, but, after a few years, even those initially opposed would be proud of such a name and patch.

Rich knew a graphic artist who drew up the patch I envisioned with a sneaky black cat reaching out to swipe the Shuttle zooming through space. A few of my class loved it. The majority hated it. Pilot Rick Searfoss started an alternate patch and name effort. The class overwhelmingly voted in favor of Rick's not-gross patch and name. But the core group that wanted the cat and the name Hairballs went ahead and created the patch anyway, and we called ourselves Hairballs. Bill was right. After a while, even the most adamantly anti-Hairball folks embraced both the rebel patch and nickname. We were all Hairballs from then on. Ellen Ochoa, who was the biggest hater of the name, had a wooden plaque made with a cat and astronaut logo, and "Prime Hairball" painted on it. It was passed around over the ensuing years to the office of the next Hairball to fly.

Q: WHAT KIND OF PHYSICAL TRAINING DID YOU DO?

OF COURSE, WE WERE expected to stay in good physical shape. For that, we had an astronaut gym on campus, with exercise equipment, a running track, racquetball courts, a half-court basketball hoop setup, and lockers and showers. We could use these facilities anytime.

Also, the first part of astronaut training was survival. Since we were going to be operational crewmembers, it was important to NASA that we trained to act as pilots or copilots in the NASA jets. We flew Northrop T-38 Talons that had been in use since the 1950s. The jets could fly at supersonic speeds and were configured for training, testing, and travel. For those of us who were already pilots, we were happy for the flight time. Those without a flying background would learn how to operate in the flight environment. We needed to think and react calmly and professionally in real time and follow a checklist. Flying jets was a great way to train for spaceflight.

We got custom-made flight helmets and oxygen masks, and were fitted for parachutes. Then we traveled to the Air Force Survival Training school at Fairchild Air Force Base in Washington state. Trainers took us parasailing to practice parachute landings without getting hurt. They dragged us behind a truck and lifted us into the air. When I was up high enough, I cut away from the truck and floated down to a bumpy landing. For some, falling through the air was a blast, but I was more into the mechanisms of flight, not the physical thrill. I'd always been like that. As a kid, I was decent at sports—but give me an old radio to take apart and put together and I was set; I found joy in how things work.

When I was nineteen, my fraternity at UC Berkeley decided to go skydiving. (I made the mistake of telling my mother, and she freaked out. *I should have told her after I'd done it, duh.*) Before we flew, we spent the day training to jump off platforms and learning how to land with a parachute. The trainer went over all the different emergency scenarios and taught us how to activate our reserve parachutes if needed. Then we went up in the airplanes and we jumped out. I didn't find it particularly scary or exciting. I was more fascinated with the airplane than jumping out of it. The experience reinforced my desire to fly airplanes.

The courses we went through at the Air Force base were watered-down versions of what the military officers and enlisted men and women were taught in SERE school, an acronym for survive, evade, resist, escape. Those guys were dumped in the forest and expected to fend for themselves, to trap rabbits or eat bugs for weeks at a time. Basically, they were hunted by other military guys playing the "enemy" who eventually captured them as POWs, put them in boxes, and put them under duress. Sounds fun.

All the while, my astronaut class was out in the field learning how to make animal traps before taking a lunch break of military Meals Ready-to-Eat (MREs). We didn't suffer. At the end of the week, we were all a few pounds chunkier. The military guys came back skinny and lay in the hallways of our shared dormitories. They didn't say a word to us. They were exhausted. They'd been roughed up. Someone ordered about fifty pizzas from Domino's. They were just *starving.* We felt a little bad about that.

In Pensacola, Florida, on a beautiful fall day with calm seas, we learned from Navy trainers how to parasail into the ocean, how to cut away our parachutes, deploy our raft, and get into it (not as easy as you may think). One of my classmates, Bill Gregory, got stung by a bunch of jellyfish and came screaming out of the water—red lines crisscrossed his legs. But he was our only casualty; it was mostly fun: beautiful nature, sandy beach, nice weather, good food, and our shared camaraderie.

Part of training is to develop a group dynamic, a team spirit. There were seven pilots and sixteen mission specialists. The class had the most women to date—five—in any class before. An Army pilot, an Air Force pilot, an Air Force flight test engineer, and two were civilian engineers. We all got along well for the most part. There weren't many points of friction, and we had fun teasing each other. Susan Helms, an Air Force Academy grad in aerospace engineering, anointed me and four other male astronauts with a sub-team nickname.

"I'm going to call you guys the BABAs," she said, laughing. "Bad Attitude Boy Astronauts." (We may have had a little attitude.)

Bill Gregory was one of the BABAs, and the one of us with the "worst attitude."

"Susan, you and your female cohort are now designated as the WAWAs—the Women's Association of Whining Astronauts."

It was 1990. There were no hurt feelings. We all thought it was hilarious.

Q: WHAT WAS THE CLASSROOM AND SIMULATOR TRAINING LIKE?

BACK IN HOUSTON AFTER survival training, our days were long. We had classwork, lectures from specialists who taught us about Space Shuttle systems and operational procedures. We spent time in the simulators working through "nominal" operations, which is what you do when things are fine, and "off nominal" operations, which is when things are screwed up, and then we also practiced emergency procedures.

The first time in the Space Shuttle simulator was so exciting. I climbed in and strapped into my five-point seat harness. I put my headsets on so I could talk to the other astronauts and the training team. The door was closed and locked. Then, after the commander gave his "go," the cabin tilted ninety degrees until we were laying on our backs. We went through the prelaunch sequence, and then at T-0, the cabin started vibrating and the simulated sound of the engines filled our ears.

We were following along on our checklist when suddenly, the master caution alarm blared. Lights lit up on the caution and warning matrix. Computer alerts appeared on the screens. In moments, we were deep into the emergency checklist working through all the problems. *What a rush!* Sometimes you do "die" in the simulator, but that just reinforces how important it is to do everything right on flight day. During generic training, we swap seats and roles and train with everyone on the team. Once you are assigned to a mission and position, you train for that specific role.

After a year, we Hairballs were ready to graduate. There is not a lot of fanfare at NASA, but we had this sort of informal class ceremony at a local restaurant, and they gave each of us a silver pin. The lapel pin was designed by the original seven astronauts. It's a star with three trailing rays of light surrounded by a ring. The design was unveiled in 1964 and written up in the Space News Roundup, a publication of what was then NASA's Manned Spacecraft Center (Johnson Space Center now). "NASA's twenty-nine astronauts are wearing a new emblem, unofficially signaling the unity of the Mercury-Gemini-Apollo flight teams," the article said. "The design shows a trio of trajectories merging in infinite space, capped by a bright shining star and encircled by an elliptical wreath denoting orbital flight." It was nice, but no one wore it because it meant that you hadn't yet been to space. It was like a cub scout pin when you were dying to be an eagle scout.

What we wanted was that same pin but in gold, the official proof that you flew on your first mission. A gold pin flies with you on your first mission, and it's presented to you formally after you get back. What's funny is that while the gold pin is free, you had to pay $15 for the silver pin. *Some NASA bureaucrat must have gotten it backward!*

Q: WHAT HAPPENS WHEN YOU GET CALLED FOR A MISSION?

THE SECOND PHASE OF astronaut training begins when you're assigned to a crew. Since my first mission was a science one, the payload crew was assigned first. The payload crew is comprised of all the astronauts that will fly the mission together (and do the experiments)—minus the commander, pilot, and flight engineer (mission specialist #2) who would be added later. I got assigned to a crew in 1992. We were STS-65. STS stands for Space Transportation System, the official name of the Space Shuttle program. It would be the 63rd Space Shuttle flight in the program and the 17th flight for Space Shuttle *Columbia*. We were set to fly in 1994 on an international science mission. As one of the last in my class to get assigned a mission, I was excited—and relieved. It was very rare, but a few times in history, there were astronauts who didn't make the cut.

The crew was comprised of people from different astronaut selection groups with some newbies and some veterans. My crew had four mission specialists: Veterans of two missions, Richard Hieb and Carl Waltz (one of my classmates on his second flight), and two NASA first-timers, Don Thomas and me. Don and I became good friends. We'd bonded during our interview when it turned out he'd been working as an aerospace engineer at JSC and had met and worked with my dad. At the time, my dad did consulting work for NASA as part of his job at Lawerence Livermore. We would also have a payload specialist on our flight, a female Japanese astronaut named Chiaki Mukai. Chiaki would be the first ever Japanese woman to go to space. Chiaki was a friendly and amiable person, smart and educated, and a hero to her country.

The crew moved into an office together, and our payload commander, Rick Hieb, got everyone together for our first meeting. All crews share an office that can accommodate up to seven people. We'd sit together to bond, share training notes, and discuss mission issues.

"This is a research flight," Hieb said. "You guys are going to start learning about all the science before we start training on how to operate the Shuttle."

"Leroy . . . You're mission specialist three. You'll be conducting research experiments and have flight crew responsibilities on the Shuttle. We all have to get up to speed on the research work and understand the underlying science and goals."

For scientists around the world, getting an experiment on the Space Shuttle was a big deal. Since gravity comes from the immense mass of the Earth, there is no place on the planet to do experiments in microgravity—except for a few seconds in a drop tower or zero-g airplane flight. The limited slots available for experiments are highly coveted. In a lengthy selection process, NASA puts out a call for proposals years before an actual mission. The proposals go through intense peer review, and projects are considered by NASA as to whether they mesh with the agency's overall mission.

The mission was the second "International Microgravity Laboratory," but we were the eleventh mission that would put the European-built Spacelab Long Module to use. The cylindrical craft was built by the European Space Agency and had a fully functioning lab space that fit about six astronauts. It also had platforms outside the module where experiments could be done in the vacuum of space.

For the next six months, we traveled to meet with the research teams that were selected to have their experiments done in space. We needed to know how to conduct their research as meticulously as if it was done in their own labs. We went all around the US, to Canada, Japan, Norway, France, the Netherlands, Germany, and Italy to meet the different researchers whose experiments were going to be on the mission. We visited the manufacturers who were developing and building the experimental hardware that would fly in the Spacelab. We helped develop the procedures to perform the experiments. We asked a million questions. *What are the things that can go wrong? What do we do in those cases to try to preserve some data and get some useful results for you?* In the end, our mission would carry the dreams and hard work of six science teams from around the world.

Many of the experiments were biological and material science based. We would have Japanese red-bellied newts onboard, along with goldfish and "medaka" fish. These animals all have balance systems (inner ear) like humans, and we were going to study their adaptation to space. On the materials side, we would have a furnace where different metal alloys could be melted and then cooled again to see what different grain structures would form in the absence of gravity. There were many other experiments, but these two were the most interesting to me.

It was an intense time. We'd spend two weeks in Europe, then two weeks in Houston, two weeks in Canada and Japan, then two weeks back in Houston, two weeks back in Europe. We were bleary-eyed and jet-lagged. It was tough doing all this traveling in economy class, contorted into tight seats, knees grazing the seatback of the person in front. I tried to sleep as best I could and hoped my brain would log on when we arrived in a lab to learn a bunch of technical stuff shortly after landing.

Q: DID EVERYONE ON THE CREW GET ALONG?

WITH ALL THE TRAVEL, learning a ton of new science and experiments, and constantly being jet-lagged, there was some drama. We weren't all hunky-dory with each other, and there were personality conflicts. Don and I were both PhDs, and Chiaki Mukai was a medical doctor. The three of us had a ton of education and research experience. Rick Hieb was the payload commander. He had a master's degree in aerospace engineering, and he'd flown before. But for some reason, it felt like he came at us with an attitude. I thought maybe he had a chip on his shoulder because he didn't have the same research background the rest of us had. He was also in his first leadership position at NASA, and I think he felt insecure about that.

"You know, you guys really need to come up to speed on all of the science, to a master's level at least," he told us.

In my opinion, that was unrealistic and unnecessary. We needed to perform the experiments and understand the objectives, not be the actual scientist.

One day Rick said, "I do have an honorary PhD, you know. Does that count?"

An emphatic "NO!" came from Don before I could answer. "I worked too hard for mine," he told me later with a chuckle. I agreed, and we sometimes still laugh about that moment.

Rick left Chiaki alone, maybe because they would be on the same "red" shift together. Don and I were on the "blue" shift. Our two teams would operate the Spacelab 24/7 on two opposite shifts.

The stress caused Don and me to have a few more beers in the evenings than we usually would have when we were traveling. Our backup payload specialist was a chain-smoking French astronaut named Jean-Jacques Favier. We bonded with him and started bumming his cigarettes and drinking beer together for a little coping time.

Q: HOW DID YOU TRAIN WITH THE WHOLE CREW?

AFTER SIX MONTHS, the commander, Bob Cabana, and the pilot, James Halsell, one of my classmates, were assigned to our mission, along with Carl Walz (also a classmate) who would be mission specialist #2, and we all started training together. My role was a backup to the flight engineer, and to help the two front seaters. I would be on the flight deck for the launch (what we called "going uphill") in the right seat behind the pilot.

I trained to assist with malfunction procedures, making sure that we'd all be safe. We did dozens of simulations in the Space Shuttle simulator, going through all the emergency procedures and reconfigurations. We trained on robotic arm operations. We trained for the spacewalks in the pool. We were drinking from a firehose, which is what all spaceflight training is like—trying to retain as much as you can, then getting refresher training. You basically spend your whole astronaut life training for those few glorious moments when you get to fly through space.

Q12: WHAT WAS YOUR BIGGEST DISAPPOINTMENT AS AN ASTRONAUT?

SINCE GALILEO FOCUSED HIS DIY telescope to see craters on our Moon in 1608, astronomers have worked to get better and bigger views of celestial bodies. Originally, telescopes were made with thick glass lenses that were heavy and difficult to handle the bigger they got. Astronomers started using mirrors made of thin, slightly concave glass to get larger lenses without the weight. They work like this: The primary mirror sees the light and bounces it to a second, smaller mirror that focuses it for the human eye. So with a big enough space telescope mirror, we can see the rings of Saturn, the stripes on Jupiter, and detailed outer galaxies.

In March 1990, the largest and most powerful telescope ever built, named after the great astronomer Edwin Hubble, was set for launch. It was sure to give us an up-close look at things we could never dream of visiting in a human lifetime.

A few years before that, I was working as an engineer for Hexcel Corporation on a project to make a new type of reflector for future space telescopes. Our hard-charging lead engineer at JPL was an MIT grad named Paul McElroy. "You're doing important work, guys," he'd say. "We're going to change the way that future space telescopes are built!" McElroy made us feel lucky and special to be working on the program, and it pumped our nerdy egos. We started calling ourselves the "Young Studs." We even designated our first prototype as "YS-1." It was a great first gig out of school. I was working for a real company and making a living as an engineer. My parents were happy. They'd done their job.

When I was still working at Hexcel on the JPL telescope project, my Hexcel colleagues and I went to a workshop organized by JPL. A guy from PerkinElmer was on the slate to speak. PerkinElmer was founded by two New Jersey astronomers in the 1930s and is still one of the foremost companies working on telescopes. I was excited to hear their research, but the presenter's attitude was immediately off-putting. He tossed plastic slides onto an overhead projector and opined on how much better PerkinElmer was than everyone else. *PerkinElmer this . . . PerkinElmer that. We're so great, blah blah blah.* And the thing he was most proud of?

"We made the primary mirror of the Hubble Space Telescope."

As pompous as this guy was, I had to admit, this was a flex. Every space contracting company wanted to be part of Hubble. And the main lens—a 2.4-meter-wide concave mirror—was Hubble's heart and soul. With this perfect instrument, Hubble could give us the most detailed look at the galaxy that humans had ever seen. It was sure to fire up an

America lukewarm about space since the end of the Moon missions.

Hubble launched on the Space Shuttle *Discovery* in April 1990, seven years late and a billion dollars over budget. The first images came back in June . . . fuzzy. Sure, they were never-before-seen celestial images, but NASA wasn't going to wow Middle America with spiraling cotton balls. When engineers looked for why they weren't getting the super high-res photos promised—ten billion times sharper than what the naked eye could see—they found the culprit. The primary mirror from PerkinElmer was flawed. It was polished incorrectly by the thickness of a fiftieth of a human hair. That tiny screwup turned into a massive blunder that threw the images askew.

NASA was sick. Congress was pissed. Maryland Senator Barbara Mikulski—an astronomy superfan—called the myopic space machine a "Techno Turkey." Hubble was now a $1.6 billion telescope orbiting the Earth with vision akin to opening your eyes in the swimming pool.

My fellow ASCANS (short for Astronaut Candidates, and what they called us astronauts in training) and I were shocked.

"God, how did they let that happen?" I wondered out loud.

"I heard that they skipped a critical test," one guy said. "They didn't do an end-to-end." An "end-to-end" test is when you put all the components of the spacecraft together to make sure everything is working the way it should.

"Yeah, they decided that since they'd done all those tests on the component level, there was no need to do the whole end-to-end test with everything together," astronaut John Young said, shaking his head.

All of us had some background in engineering projects and/or flight tests. We knew that testing is everything. NASA and PerkinElmer would have certainly caught the problem if they had done an end-to-end test.

NASA blamed PerkinElmer. PerkinElmer blamed NASA. In 1982, NASA had gone to Connecticut and told them to hand over the merchandise. It was not ready, PerkinElmer claimed. They wanted more time to fix it since the tests were showing something off. "We were told to stop," Bud Rigby, the project manager, later said. "They shut us down in four or five days."

The mirror got shipped across the country to Lockheed Martin, where a whole different team of engineers put the telescope together. "It was like being a surgeon and being two-thirds of the way through an operation," said PerkinElmer engineer Lou Montagnino in an interview years later. "Then, someone walks into the operating room and tells you, 'You guys are done. We'll close it up.'"

I knew that NASA was not above cutting corners, and it came out that skipping the test was an agency decision to save money. In hindsight, it's easy to say, *What a bunch of idiots. Why didn't they do this test?* But if they'd gotten away with it, then NASA management would have boasted that they'd successfully run the mission and had saved money while doing it. It's hard to be too critical, but I was kind of in disbelief that NASA let it happen.

NASA was a struggling agency around the time I was selected to be an astronaut. Richard Truly, the Administrator of NASA since 1989, had a CV that read like Captain America: Fighter Pilot, Astronaut, Navy Admiral. But after the *Challenger* explosion and the general public's disinterest in space, Truly battled with political leaders and often had a hard time getting decent budgets.

It wasn't all bad news. President George H. W. Bush had announced the Space Exploration Initiative (SEI) just a year earlier, on the 20th anniversary of the Apollo 11 Moon landing. The goal was to land astronauts on Mars by July 20, 2019, the 50th anniversary. Bush asked Congress for a 24 percent increase in NASA's budget, and the program had the support of the Budget Director, Richard Darman. Things were looking up.

We all knew that we had one of the best jobs in the world, and although NASA experienced major setbacks over the prior few years, it was an exciting time to be an ASCAN. It seemed that some of us might actually have a shot at going to Mars, or perhaps at least the Moon, under the new program. Like the twelve ASCAN groups before us, we tried to excel and stand out as individuals to get assigned our first missions. It's sort of a contest. You get to training, then you basically wait and hope. You don't know when or where you'll be when the call comes.

One of my favorite things in training was learning how to perform Extravehicular Activity (EVAs)—or spacewalking. We all wanted to fly into space, that was a given. But most of us wanted to leave the confines of a spaceship and *walk* in space too.

In the 1970s, there was a full auto shop at my public high school, complete with a hydraulic lift, power tools, and air tools. I learned to do tune-ups, brake and clutch jobs, pack wheel bearings, you name it. I always loved building and fixing things and using tools. The experience not only made me a better engineer, it also taught me the skills to be a better EVA astronaut. Fixing cars gives you contortionist skills to work in small spaces, and the finesse to get the right wrenches into tight places.

By the time I put on the spacesuit and was lowered into the water tank for training, only around fifty people had ever left a spacecraft and entered the vastness of our universe. Obviously, with very few opportunities, EVA was not an easy gig to land at NASA or any other space agency. Walking in space is like a final frontier for humans; there is no place farther away from our existence. There is not a more elite club to belong to. I was determined to get on that list.

In my first year of astronaut training, in 1991, there were no big NASA projects that would need astronauts to do EVA. But the following year, we got word that the fix for Hubble was ready. The Hubble Space Telescope repair mission would give the telescope some glasses to correct its bad vision, install a new advanced camera, and make a few other tweaks. It would take a team of four spacewalkers.

This was the hot mission we all wanted. I knew I had a chance. Every astronaut did training runs in the water tank. One of the rites of passage was using special tools on the Space Shuttle mockup to secure the payload bay latches in case of failure. If the payload bay was not securely latched, the 3500-degree plasma could flow in on reentry and cause a catastrophic failure, exploding the Space Shuttle and the astronauts aboard.

From the beginning of our training, I showed an ability for EVA. I used the same skills I honed way back in my dad's garage. I was always tinkering with things, using tools to make model boats, airplanes, and cars. I loved working with my hands. And my high school auto shop was like a graduate school for these skills.

When I first put on a spacesuit, it felt weird. It's heavy, and once it's pressurized, it takes strength to move your arms and your fingers. The visibility is limited by the helmet, so you're already feeling awkward when they drop you into the giant water tank. To simulate floating in zero gravity, divers come and add weights in different pockets until you're neutrally buoyant. The metal tools we use in space are heavy on Earth, and even in water, so you've got to have the strength to hoist them up to the various places on the mockups.

During the so-called "generic" training for EVA, you practice several tasks, including securing the payload bay doors and latches. In case there's a problem with the payload bay door latch, you use tools that are designed to manually crank the doors down to make sure they're sealed for entry. You've got to find a way to steady yourself and be really careful with the tools. If you "drop" them, they fall toward the bottom of the pool (they are tethered to the spacesuit), and you've got to start over. A lot of people can't do it because you simply run out of arm strength. A fair number of people struggle, but I had no trouble getting it done.

Because I showed ability in the spacesuit, the astronaut in charge of spacewalk development sent me to Alabama. The mission to repair Hubble was managed out of Goddard Space Flight Center in Maryland, with development and training out of the Marshall Space Flight Center out in Huntsville, Alabama. We used the giant cylindrical water tank at Marshall for developing the steps, procedures, and tools that would be used to fix the Hubble Space Telescope in space. There was a full-size mockup of the telescope in the water tank, and a working robotic arm that mimicked the one on the Shuttle. We spent hours in our spacesuits in the water, trying different techniques and positions while evaluating the tools the astronauts would use in space.

We went through different ways to install the corrective lenses for the telescope, as well as for removing the Wide Field

Planetary Camera and installing a new, advanced version. The runs were successful, and fixing and updating Hubble looked very promising. During team debriefs, Don Thomas, my teammate and fellow ASCAN, and I received compliments from both the Hubble team as well as from the EVA team from Houston.

I was riding high. I felt like maybe Don and I had a shot at this mission.

Several months later, a senior NASA official pulled me aside.

"Hey, Leroy, you've been doing a great job on EVA training and the HST development runs. So get ready to go do a bunch of spacewalks."

I knew there were no other spacewalking missions on the horizon except the Hubble repair flight. I was going to get to walk in space on my first mission and get Hubble back to 20/20 vision!

"Yes, sir! Thank you, sir!" I said enthusiastically. I couldn't get the smile off my face.

I knew he wasn't supposed to tell me that ahead of the official selection, but I was glad he did. I was so fired up.

To do an EVA mission, you have to be diligent about getting to the gym and working out. I was already in good shape, running and doing resistance training pretty much every work day. My classmate Eileen Collins, an Air Force test pilot, got the same down-low message that she would be the mission's pilot. Carl Meade, from a more senior class, was also told, like me, that he would be assigned and do EVAs. Eileen and I talked about it together. We were close because we were from the same astronaut class.

"This is going to be great" she said.

"Yes," I agreed. "What a great mission!"

Not long after, I started hearing rumblings that George Abbey, the "Godfather" of NASA human spaceflight and Special Assistant to NASA Administrator Dan Goldin, was quietly pushing back. He had been "kicked upstairs" out of the Johnson Space Center to a desk job at NASA headquarters in Washington, DC, just before my selection. My astronaut class was the first in the Shuttle era that was not selected by him. Abbey wasn't happy about Eileen, Carl, and me, because he felt that rookies shouldn't be on this flight (Carl had flown in space before but had not done EVA). The top brass was still under a lot of heat for the Hubble mess.

A few weeks later, Dan Brandenstein, the chief of the astronaut office, called an all-hands meeting. It's what they do when they're going to announce a crew.

Huh? I hadn't been officially notified that I was in, and they tell each astronaut individually before they make the formal announcement. *But I'm on this crew!*

There was a pit in my stomach.

Brandenstein put up the view graph. LEROY CHIAO wasn't on the list.

What the hell?

Eileen wasn't either. Neither was Carl. They hadn't even told us. I was bummed, but I was trying to have a positive attitude about it. *Well, I'm still an astronaut. I'm still going to fly.* I went from what would have been the coolest flight—a high-profile mission to fix the Hubble Space Telescope, to later being assigned to a research mission on the Space Shuttle. It stung. I went from being the guy in my class to get the coolest mission, to one of the last guys in my class to fly. But, in the big picture, I couldn't complain.

Q: DID IT FINALLY GET REPAIRED?

VETERAN SPACEWALKERS JEFF HOFFMAN and Kathy Thornton replaced Carl and me, and veteran pilot Ken Bowersox replaced Eileen. Watching them train to fix Hubble was tough. But when I saw the team in space, I couldn't help but be moved by the sheer poetry of the mission. Like worker bees around their regal queen, four astronauts spent about thirty hours tending to and tinkering around that majestic space telescope. The mission glided to its finish with no major hiccups.

On December 18, 2003, the first image came from the new and improved Hubble. It was a star named M100—a crystal clear diamond surrounded by a swirling pinwheel of blue sparkles. Maryland Senator Barbara Mikulski got a sneak preview before the press conference where NASA had set up a before and after photo. "My God, it's like putting my glasses on!" she said.

I was part of the bigger NASA, so I was happy for Hubble and everyone on Earth who could now see close-up photos of space. But for me, photos weren't enough. I still needed to be there in person.

Chiao family portrait in 1973.

PHOTO COURTESY OF LEROY CHIAO

Leroy Chiao with his paternal grandmother at his graduation from the University of California.

PHOTO COURTESY OF LEROY CHIAO

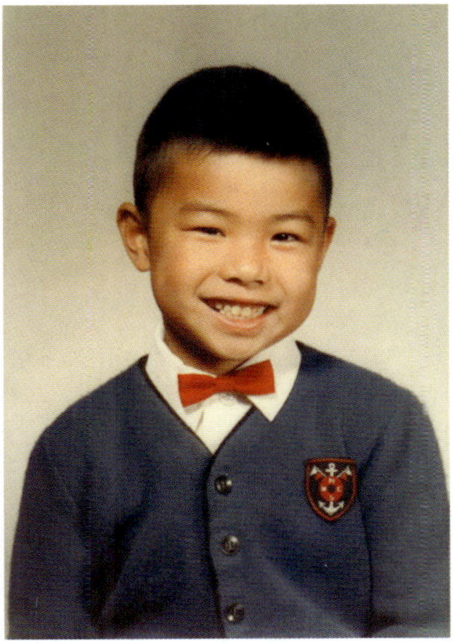

Leroy Chiao's second-grade school portrait. Vista Grande Elementary School, Danville, CA.

PHOTO COURTESY OF LEROY CHIAO

Leroy Chiao at age three, Milwaukee, WI.

PHOTO COURTESY OF LEROY CHIAO

Leroy Chiao during EVA, on Space Shuttle mission STS-92.

PHOTO COURTESY OF NASA

Yuri Shargin, Salizhan Sharipov, and Leroy Chiao after a checkout exercise of their Soyuz TMA-5 spacecraft.

PHOTO COURTESY OF NASA

Leroy Chiao became the first American allowed to visit the Astronaut Center of China in 2006. This photograph is from one of many subsequent trips to the center. Chiao's good friend David Warden is fourth from the right.

PHOTO COURTESY OF LEROY CHIAO

Launch of Soyuz TMA-5. We are on our way to the ISS (October 14, 2004)!

PHOTO COURTESY OF NASA

Leroy Chiao before an EVA training session (1991). Mission Specialist astronauts go through a generic EVA training program, where they are evaluated on their skills and suitability for EVA missions. In this photograph Chiao is about to be lowered into the water to begin one such session.

PHOTO COURTESY OF NASA

An official NASA portrait of Leroy Chiao in an EVA spacesuit (1997).

PHOTO COURTESY OF NASA

Don Thomas is an astronaut classmate of Leroy Chiao. They flew their first space mission together on STS-65 in July 1994. Both astronauts were inspired by the Apollo 11 Moon landing, and STS-65 coincidentally flew during the 25th anniversary of Apollo 11.

PHOTO COURTESY OF NASA

Debbie Gary was the first woman to fly on a professional aerobatic team. Over the years, she earned many honors and awards. This photograph was taken at a private airstrip, from which both she and Leroy Chiao operate their airplanes.

PHOTO COURTESY OF ERIC FORSYTHE

Jaime Forero worked for nearly four decades at NASA, primarily in the Vehicle Integration Test Team (VITT). He became good friends with Leroy Chiao and the two worked together often over the years.

PHOTO COURTESY OF NASA

Leroy Chiao undergoing Space Shuttle Water Survival Training. Certain abort modes of the Space Shuttle involved parachuting into the ocean. Astronauts are trained in sea survival for these possible contingencies.

PHOTO COURTESY OF NASA

The Moon captured from orbit aboard ISS.

LEROY CHIAO, COURTESY OF NASA

Leroy Chiao and Roland Nedelkovich, his best friend in Houston (1993). The two met in 1991 when they were both working in the Shuttle Avionics and Integration Lab (SAIL).

PHOTO COURTESY OF LEROY CHIAO

Leroy Chiao and Rich Clifford, his best friend in Astronaut Group 13 (1995).

PHOTO COURTESY OF LEROY CHIAO

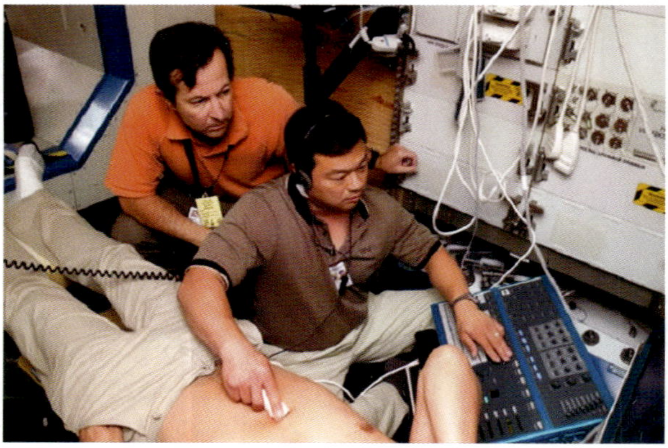

Leroy Chiao and Dr. Scott Dulchavsky during ISS training (2004). The two became good friends during training.

PHOTO COURTESY OF NASA

Leroy Chiao and his classmates are briefed on a T-38 land survival course (1990). In preparation for flying as a crewmember in the NASA T-38 jets, all Astronaut Candidates (ASCANs) undergo land and sea survival training through the US military.

PHOTO COURTESY OF NASA

Class photo of NASA Astronaut Group 13 (1990).

PHOTO COURTESY OF NASA

Leroy Chiao and Steve Valles, his good friend and college fraternity brother (1992). Valles and Chiao both pledged the Phi Kappa Tau fraternity at the University of California in 1978. They have remained good friends over the years.

PHOTO COURTESY OF LEROY CHIAO

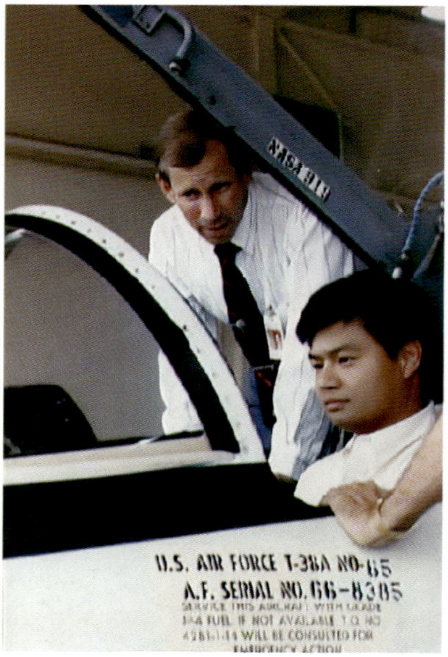

Ken Cockrell briefs Leroy Chiao on the T-38 Jet (1990). Cockrell was a career Naval Aviator and a NASA staff pilot before being selected to be a member of NASA Astronaut Group 13. He was class leader and shared an office with Chiao during their first year of training together. They remain good friends and are neighbors in the same flying community.

PHOTO COURTESY OF NASA

Leroy Chiao and Salizhan Sharipov begin an EVA training session in the Hydrolab in Star City, Russia (2004). They executed two Russian EVAs during ISS-10.

PHOTO COURTESY OF LEROY CHIAO

Leroy Chiao inside a Soyuz simulator in Star City (2004). All spacecraft require many hours of training based mostly in high-fidelity simulators. In this photograph, Chiao is about to begin one such training session.

PHOTO COURTESY OF LEROY CHIAO

Leroy Chiao and Jami before an astronaut reunion event in Houston (2022)

PHOTO COURTESY OF LEROY CHIAO

This is the first official NASA portrait of Leroy Chiao, taken shortly after his arrival at the space center as a member of NASA Astronaut Group 13 (July 1990).

PHOTO COURTESY OF NASA

Leroy Chiao's children, Henry and Caroline, at an Easter egg hunt (2014).

PHOTO COURTESY OF LEROY CHIAO

Progress 17P resupply ship, station-keeping at fifty meters from ISS.

LEROY CHIAO, COURTESY OF NASA

Q13: WHAT WAS IT LIKE LAUNCHING TO SPACE FOR THE FIRST TIME?

I WAS NERVOUS, but not for the reason you might think. I wasn't scared about dying, because we all understood the risks we were taking. Part of the job was sitting on top of a rocket with huge propellant tanks feeding three liquid engines through high-speed turbo pumps, and two huge solid rocket boosters attached to the sides. Everything had to work almost perfectly. Of course there's risk. I'd thought about it for years. I'd accepted the "risk versus reward" calculation.

What did stress me out the closer we got to our July 8, 1994, launch? I was afraid something was going to keep me from *finally* going to space. *What if some health problem knocks me out medically? What if I get in a car crash?* I was thirty-three and in perfect health, but as we got closer to launch date, I drove around Houston like a grandma, took loads of vitamin C, and always looked both ways before crossing the street.

The week before launch, we went into quarantine to minimize the chance of us getting sick. Everybody who came in to see us had to be checked out. Flight surgeons gave cursory physicals to make sure visitors didn't have a cough or a cold or anything contagious. Because I was single with no kids, I could stay at my house at night. My girlfriend, Alece, had been medically cleared, so she was allowed to visit. I couldn't go to her house or anywhere public or let anyone else in my house who wasn't cleared. Astronauts with children under eighteen couldn't stay at home since kids are basically germ factories. The first half of the quarantine was at Johnson Space Center then we would finish quarantine in Florida at the Kennedy Space Center—the site of all Space Shuttle launches.

Q: WHAT WAS QUARANTINE LIKE?

THOSE WERE QUIET DAYS. In Houston, we had a basic dorm with a kitchen and office area, where I went over my training notes to keep my brain fresh. We spent a lot of time signing crew photographs with black Sharpies. Each mission takes a group photo to give to the flight controllers, the training team—all the people at NASA that support the mission. The photo was the seven of us in our orange flight suits. I signed over my image on the left next to the American flag and made a stack to pass to the next guy. Even with my truncated signature to save time—sometimes only a squiggly *La* Ch with a dot above an invisible *i*—my hand cramped up and I got sick of smelling Sharpie fumes.

It was a little like house arrest except for the services we enjoyed. We had every meal cooked for us. There were a lot of cookies and chips and snacks that I tried to avoid. The members of the Vehicle Integration Test Team (VITT) spent most of their time monitoring the processing of the vehicles and payloads and work issues. They were our eyes and ears and took care of everything we couldn't do for ourselves in quarantine. They even brought beer for us, which I didn't mind at all. Two of my best friends at NASA were in the VITT: Jaime Forero, a long-time "VITT-guy," who worked all three of my Shuttle missions, and Roland Nedelkovich, who I met while we were both working in the Shuttle Avionics Integration Laboratory (SAIL) during my first technical assignment as a newbie astronaut. We became best buddies, and then Roland moved over to the VITT after my first Shuttle mission. He worked my second mission and was the VITT lead on my third.

Nobody was scrolling on their phone back then, so we passed the time studying our notes on flight procedures, the experiments we'd be working on in the Spacelab, and creating our crew notebooks. These were small blank notebooks where we would write tips and reminders to ourselves for everything we would be doing during the mission. To break it up, I read novels by Tom Clancy, Michael Crichton, and Arthur C. Clarke. We had *Sports Illustrated* and *Time* magazines, and movies on VHS like *Caddy Shack*, *The Blues Brothers*, and *Stripes*. If there were any specific titles we wanted to watch that were not in the library, the VITT guys would rent them for us from Blockbuster. Anything Monty Python was a personal favorite and became a crew favorite on my third mission.

Q: WHAT WERE THE LAST FEW DAYS BEFORE LAUNCH LIKE?

AT T MINUS FOUR DAYS, we flew down to Kennedy Space Center in NASA T-38 jets. Astronauts are required to log a number of hours in the jets each month, to maintain operational readiness in the flight environment. These sleek jets carried a crew of two in tandem. The military-trained pilot astronauts flew from up front, and most mission specialist astronauts like me flew in the rear cockpit as copilots.

A few years before, I had earned a private pilot license while a graduate student in Santa Barbara. I continued my training after graduation and earned an instrument rating a few years later. Things happen a lot quicker in a jet compared to a light airplane, and the controls are also a lot more sensitive. It took practice to be able to hold altitude and work quickly enough to keep up. After the jet training and practice, Cessnas and other airplanes I rented for personal flying seemed so slow. I became a much better pilot thanks to flying the jets.

In about two-thirds the time it would take a commercial jet, we passed over the aquamarine Indian River separating mainland Florida and Merritt Island and touched down on the 15,000-foot runway at NASA's Shuttle Landing Facility at the Kennedy Space Center.

Q: WHY NOT JUST LAUNCH FROM HOUSTON INSTEAD OF TRAVELING TO FLORIDA?

EAST OF ORLANDO, FLORIDA, lies what's referred to as the Space Coast, and there are good reasons to launch rockets from there. You might think that volatile weather and hurricane season in south Florida would make it hard to launch spacecraft. And you're not wrong. Many flights get scrubbed because of weather. But the reason it works well goes back to the physics of flight.

Since the Earth spins counterclockwise (viewed from the north, or top of our globe, in what's called an inertial reference frame), the rotational speed of the Earth at the equator is the highest (around 1000 mph) and diminishes as your launch site is moved toward either pole. This is "free" speed, which means you can carry less propellant to get to the orbital speed of around 17,500 mph. So, it makes sense to launch in an eastward direction, and from as close to the equator as you can.

Cape Canaveral—which is on a barrier island just southeast from Kennedy Space Center—sits at a latitude of about 28.5 degrees, so it's a good place to take advantage of this free speed. Another good reason to choose the coast of Florida? Rockets launch to the east over the Atlantic Ocean. You don't want to launch a giant metal object containing burning fuel above where people live—*for obvious reasons.*

If you want to reach a target vehicle in orbit, like the ISS, a satellite, or a space telescope, you need to launch into the same orbital plane—like getting into the same groove on an old-fashion vinyl record. To get to higher latitude orbital planes, you need to steer your rocket into that plane right after launch. From Cape Canaveral, the Florida coastline allows a rocket to steer northward after launch into up to a 67-degree orbital plane while still staying over the ocean. Houston also sits at about the same latitude as the Cape, but any rockets launched from there would have to fly over populated areas. A due east launch from Houston would place the vehicle over Florida.

Q: DO YOU GET TO SEE YOUR FAMILY BEFORE YOU GO TO SPACE?

WE WENT INTO QUARANTINE at the Kennedy Space Center, and my girlfriend, mom, dad, sisters, and some friends came out a few days ahead to watch the launch. NASA set up events for our launch guests, and there were some events we could participate in with a select few of our guests. Some of it was like Parents' Day for college freshmen in a way.

We astronauts had to keep a distance because of quarantine, but the coolest large-group event for a select number of our guests was the Shuttle night viewing. While *Columbia* was getting ready on the launch pad at Cape Canaveral, it was hidden behind a metal structure called the Rotating Service Structure, where platforms were attached for technical teams to work on the Shuttle's exterior. On viewing night, they rotated the cover out of the way so we could see the whole Shuttle. With super bright lights on the vehicle, it was a spectacular sight—the 154-foot-tall orange oxygen tank towering over the Shuttle standing on its tail—like a giant totem pole to the space gods.

Sometimes while crews were in quarantine at Kennedy Space Center, there would be an opportunity for the crew to greet guests from across a ditch, far enough away to keep us safe from any germs.

I shouted across the ditch to my family. I was dressed in my blue flight suit on the other side with the crew.

My family waved back.

This "take your parents to work" day felt great. I knew Mom and Dad were proud of me, even though my dad would never admit it out loud.

My girlfriend, Alece, and I met when I first arrived at NASA. She'd never seemed concerned about my safety flying to space before. But as we got close to my first launch, she started getting worried.

In truth, spaceflight was safer than the early days, but failure could still be catastrophic.

During launch and ascent, we had what were called abort boundaries (which define abort zones) and we trained for abort scenarios. Certain actions would be taken within an abort zone. Once past the boundary of that zone, other actions would be appropriate in the new zone. Not all zones were considered survivable. These were called black zones. We accepted them but tried not to give them too much thought.

If something happened early on like an engine failure or a bird strike that busts a window, we would try an abort called Return To Launch Site. The procedure called for jettisoning the solid rocket boosters and trying to make a U-turn with the Shuttle main engines burning at maximum thrust. We'd be flying back through our own exhaust at Mach 4—four times the speed of sound—until we slowed down and then changed our path back toward Kennedy Space Center for a landing at the Shuttle Landing Facility. Computer models showed it was possible. We'd practiced it in the simulator and made it work. *But still, would it really work? Would the Shuttle really stay together in one piece?* If something happened later in the launch while we were in the Trans-Atlantic (TAL) abort zone, we would have enough speed to make it to Europe or North Africa. The commander and pilot could glide the Shuttle across the ocean to one of the remote landing sites. That was considered a lot more survivable than trying to fly backwards at supersonic speeds.

The only scenario we trained for that involved blowing open the overhead window was if we were stopped on the runway, and something happened that made it necessary to get out of the vehicle immediately, like a fire in the crew module. In such a case, we would blow the side hatch and the overhead window. Astronauts on the flight deck would climb out of that window and rappel down to the ground. Astronauts on the middeck would go out the side hatch and do the same. If we were in a black zone, I figured all resources would be fair game.

I didn't go into detail with Alece, but in a catastrophic failure, I was going to pull that handle. I was going to blow that window open. I'd find a way to get out of there along with as many of my buddies as possible. My parachute would open, and I'd make it down. I told myself, *I'm going to be able to do that.* Of course, the reality would probably be different. But I've always had this mentality of never giving up. I've always believed that I'd find a way out if there was the slightest chance. You have to think that way. Otherwise, why would you ever get into the beast?

Each of us had five medically cleared family members or companions, so I was able to take Alece's hand. "Look," I said. "In the worst-case scenario, I'm going to find a way out."

"I hope that's true, Leroy."

"Don't worry. It's going to be fine. If something goes wrong, I'm not going to give up. I'm going to find a way to survive."

Q: WHAT DID YOU DO THE NIGHT BEFORE LAUNCH?

T MINUS ONE DAY, we got our weather briefing. The forecast was lousy, for the Cape *and* the Trans-Atlantic abort sites. It looked like rain and thunderstorms everywhere, and the launch would probably be scrubbed. Even with the high likelihood of a canceled flight, NASA decided that we would still fill the Shuttle's External Tank ("tanking" means filling the propellant tanks in the big orange external tank with liquid oxygen and hydrogen) and go through all the prelaunch procedures. We would suit up, climb in the Shuttle, get strapped in by the closeout crew, then lay on our backs for two hours of final prelaunch procedures. After the weather abort, we figured we'd get unstrapped and go back to the crew quarters for burgers and beers and wait to hear when we'd have another shot. "What would you like for breakfast tomorrow?" The cook went around the room taking orders.

"I'll just have some toast," someone said.

"I'm going to have some cereal," said another.

Some astronauts get worried about queasiness on their first launch into space. You don't know how your body's going to react. So, if you're going to throw up, most people want their barf to be on the lower scale of grossness. I thought the opposite way. Being a superfan of astronaut history, I knew that the Apollo astronauts had steak and eggs before every launch. It was high protein and low fiber. There were no bathroom breaks for a while, so they didn't want a lot of fiber or liquid.

"You're going to eat steak and eggs?" Don Thomas, my fellow first-timer and mission specialist buddy, asked, looking at me funny.

"Yeah. If I'm going to puke . . . well . . . go big or go home!" I smiled. "Besides, that's what all the Apollo astronauts had for prelaunch breakfast." Don was also a big fan of all things astronaut from the early days.

He laughed, thought for a second, then canceled his cereal.

"I'll have what he's having! But no eggs." Don didn't eat eggs, but he was all in otherwise.

On the side of our Shuttle seats, you could clip a soft bag to hold things like your checklists and Ziploc bags of pens and clips. I hadn't barfed in the zero-g plane during training, but I did have a barf bag in the kit. *Just in case.*

To help with the motion sickness, the flight surgeons chopped up a Phenergan tablet and a Dexedrine tablet and put a half of each pill in a capsule ("Phen-Dex"). The Phenergan was an effective anti-nausea drug, but it made you sleepy. The Dexedrine was a stimulant to wake you back up.

We were launching late in the morning, about 10:30 a.m. We had dinner then went to bed, still thinking that we would have to scrub the launch due to the storms. Some of the crew took sleeping pills because the excitement and anticipation made it hard for them to sleep. I took one too and slept like a baby—at least for a while because the pills only last about four or so hours.

Q: IF WE CAN FLY AN AIRPLANE IN RAIN, WHY NOT LAUNCH THE SHUTTLE?

NASA WAS WORRIED ABOUT rain hitting the orange foam insulation on the external fuel tank and the leading edge of the wings during launch. Going fast enough, raindrops travel like bullets and can erode the foam and damage the wings. A chunk of foam can come like a missile into the Shuttle fuselage or wings. I wasn't so worried at the time, but nine years later, disaster would strike a different mission on this same shuttle I was about to fly, *Columbia*. A piece of foam would peel off during just about the worst time during launch (high speed and high pressure), accelerate to over five hundred miles an hour, and smash into the leading edge of the wing.

NASA cameras saw a big piece of foam strike the wing and break apart. Sadly, NASA engineers weren't too worried at the time. *It's like the lid off a Styrofoam cooler flew off a truck and hit your windshield. It's no big deal.* When the Shuttle re-entered the atmosphere sixteen days later, hot plasma invaded that damaged wing and burned it up from the inside. The Shuttle broke apart and the crew—who were just minutes from home—perished.

The temperature and weather parameters to clear the Shuttle for launch were conservative and complicated. Wind speed has to be within nineteen and thirty-four knots, depending on the direction. If there was any rain at all on the launch pad or in the flight path, the launch would be scrubbed. If lightning occurred withing thirty miles from the launch pad or path, the flight would be delayed or canceled. A complex system considered the type of clouds in the path, their thickness, and their temperature at certain altitudes. If those numbers didn't add up to a smooth and safe launch, the flight would get delayed.

"It's not looking like we are going to launch tomorrow," Bob Cabana, the commander, said at the briefing the night before.

My heart sank a little, but I wasn't surprised.

Q: RIGHT BEFORE LAUNCH—ARE YOU GETTING NERVOUS?

IN THE MORNING, the weather briefers were back. The forecast had improved just a little bit, but the weather at the Cape and the TAL abort sites were still no-go, or "red." Tanking was mostly complete. Everyone was excited. We joked around. We laughed while we suited up in our partial pressure suits—the orange "pumpkin suits" that you see in photos of us when we're about to take off. The bright international orange color was so if we had to bail out of the Shuttle, we'd be easier to spot in the ocean or on the ground in dense foliage. They're a *lot*, so we had suit technicians helping us put everything on.

But first . . . we put on diapers. (When the cosmonauts learned about this, they looked down their noses at us a bit.) NASA bought adult diapers, cut them open, and put more absorbent material in them. I emptied my bladder as best I could that morning and drank almost nothing. I didn't want to have my first launch experience feeling like I was on a car ride with Dad who wouldn't pull over. Once we were suited up, we laid back in easy chairs to make sure the suit felt right in the position that we'd be in "going uphill."

I always laugh about all the traditions and superstitions that the Russian cosmonauts and specialists have: champagne with the lead cosmonaut and other officials before going to the suit room, cognac shots with the chief designer just before suit-up, peeing on the tire of the bus at a stop just before getting to the launch pad (Yuri Gagarin had done it and had a successful flight, so every [male] flyer had done so too since then). But we at NASA had them too. Before every Shuttle launch, the crew commander played this card game called Possum Fargo with the Director of Flight Crew Operations. They'd play until the commander got the lowest score. It means he won, and he was leaving all the bad luck behind. Some Shuttle astronauts used to leave a bar tab open at the astronaut watering hole called "The Outpost" for good luck, implying that they would return.

A crowd came out to greet us as we got into the bus. We waved to everyone—pretty much all NASA employees—who were cheering us on.

Many would think that we'd all be serious at the point right before launch, but in reality, we were cutting up and laughing. None of us felt that there was anything to be somber about. *Yeah, we're about to get into this dangerous vehicle, but this is our jam!*

The big external tank was full of propellant, and the nature of liquid hydrogen and oxygen is that you have to keep replenishing it because these propellants are evaporating in the tanks, and the gas must be vented off in order not to over-pressurize the tanks. Clouds of vented oxygen and hydrogen were coming out of the vents, and the tank creaked and groaned with the pressure and temperature changes. The beast was alive!

We got in the elevator and rode up to the 195-foot level of the launch pad, where the "white room" is located, on the pathway to the shuttle's hatch. The closeout crew was there dressed in their clean "bunny suits." They took us one at a time; first the commander, then the pilot, then I was next.

Since the Shuttle is vertical and our seats are basically on the wall, we needed help getting into them. I wriggled my way into my seat, and the tech helped strap me in. I had my helmet on with the visor open.

They brought in Carl Walz who sat behind the commander, then the last three astronauts who were on the middeck behind us. After the last person was secured, the leader of the closeout team wished us well.

"Good luck," he said.

"Thank you."

He closed the hatch and locked us in. We fully expected to see him again in a couple of hours when they officially scrubbed the launch because of weather. We would just lay on our backs for two and a half hours, the closeout crew would come back, open the hatch, and we'd reverse the process. We'd get out of our suits, go back to crew quarters, and have a cheeseburger.

It seems unbelievable, but there was nothing to do for a lot of time, so I took a little nap. I wasn't going to be throwing any switches. I was just sitting there listening to the commander occasionally talking to the launch control center, and I dozed off.

Q: WHEN DOES IT START GETTING REAL?

AT T MINUS TWENTY minutes there was a built-in hold. That meant the countdown clock stopped, and they did a round of checks to make sure everything was okay. We were all very awake at that point, but we were still getting reports from the launch Director on the weather.

It isn't great in Florida. It's not great overseas at the abort sites, but it is improving a little bit.

At T minus nine minutes there was another hold. The clock stopped. Everything was fine. The shuttle was good to go.

But "the weather still is marginal," the launch Director said. "We're going to go down to T minus five and hold."

I was waiting . . . waiting. The anticipation was wild.

A few minutes later, the launch Director calls in.

"Cape weather is now green. TAL weather still red, but improving"

At about T minus six minutes, the launch Director called, "Cape weather green, TAL weather now green. We are not going to hold at five."

Holy crap! We're going to fly.

Adrenaline shot through my veins like jet fuel. There was no small talk, no chatter. We were silent except for radio calls between the commander and the launch Director, and when we did our communications checks. I could almost hear my heart beating.

T minus one minute thirty seconds

Commander: *Tabs in, everyone. Fans off.*

We had tabs that pulled the neck dam of the suit away from the skin to allow the fans to blow cooling air through the suit and out of those neck openings. The neck dam was necessary because the partial pressure suit also functioned as our sea survival suit, and that part kept cold seawater from getting into the suit from the neck. I tucked the tabs in the collar of my suit so my visor could seal.

T minus one minute.

Launch Director: Columbia, *close and lock your visors, initiate oxygen flow, and have a good flight.*

We all checked in that our visors were closed and that we'd switched on our breathing oxygen.

T minus seven seconds.

Commander: *Get ready for the biggest roar of your life.*

At T minus six point five seconds, the three main liquid fuel engines were lit. I was wearing a communications cap with a closed helmet, but I still heard this enormous roar, and the entire crew module shook. The whole stack (the shuttle and the rockets) pitched forward three feet with one and a half million pounds of thrust from the main engines, which were mounted at an angle from the stack.

We were still pinned to Earth because each of the solid rocket boosters had four huge bolts that held everything on the launch pad.

T minus zero.

The second I'd been waiting for all my life!

The computer issued the command to light the solid rocket boosters and simultaneously explode the bolts holding us down.

Boom!

Five and a half million additional pounds of thrust from the boosters exploded us off the launch pad. It was like somebody kicked the back of my seat. We were going one hundred miles an hour in two hundred feet as we cleared the launch tower. A little after one minute, we were flying faster than the speed of sound. The solid rocket fuel, by its nature, burned unevenly and caused a tremendous vibration. I couldn't even read the computer screens because we were shaking so much.

The pitch of the wind whistling got higher and higher because we were accelerating so fast, but the intense noise was muffled because of our helmets and communication caps. I pulled out a mirror from under my knee pad and looked through the overhead window. The vehicle had pitched over on its back shortly after liftoff (this is a planned roll and how the Shuttle flies upside down in orbit), and I could see the fire behind us and the Earth zooming away. Just *poof.*

At that point, control shifted from the launch control center at Kennedy to the Mission Control Center (MCC) in Houston.

We were roughly two and a half minutes in, at about 180,000 feet, when the solid rocket motors burned out. Very quickly, we lost five and a half million pounds of thrust. It was like someone suddenly lifted their foot off the accelerator. It doesn't matter how many times you fly in the simulator; you really don't have a feeling for this reality. I freaked out for a second.

Wow, we're slowing way down. All the engines shut down, including the three shuttle main engines. We're dead.

I looked at the engine instrument panel. All three main engines were still up, in the green.

Phew.

Then, *BOOM!*

There was a flash in the side windows as the explosive bolts jettisoned the empty solid rockets into the high atmosphere.

Silence, or at least relative silence.

We were traveling on the liquid oxygen–hydrogen engines. I felt the G-force on my chest, so I knew we were still accelerating, but I felt like I could have been sitting in a quiet room.

After about five minutes, we reached the altitude of fifty miles—what officials during the X-15 program defined as the boundary of space.

"Okay, we've got four new astronauts on board," the commander said.

I smiled big.

We all raised our visors. We were going so fast and so high up that there would be no returning anytime soon. We kept accelerating up to 17,500 miles an hour—our orbital velocity. Once in orbit, we didn't need a fuel source to keep us moving.

Then, the main engines cut off, right on time. We were instantly weightless. Our bodies relaxed and moved forward in the seats because we were released from the acceleration G's. Moments later, there was a dull thud. Explosive bolts blasted the external tank off the shuttle—a process called external tank separation, or ET Sep.

Instantly after the main engine cut off, or MECO, I was super dizzy. My inner ear told my brain that I was spinning in all three axes, but my eyes told my brain I was steady. That mismatch is like being seasick.

"Oh, wow," I said out loud.

"Okay, go easy, everyone," the commander said. "Move very slowly so you don't get nauseous."

I was floating, and it was *so* weird. I took off my helmet and my gloves, put them in my bag and unstrapped myself. I was careful not to bump into any switches or grab anything I wasn't supposed to. I was tasked to film the external tank flying away from us, so I went to get the camcorder parts.

I had a few moments to look at the Earth. It was like a dream. I had worked so hard to be here. I got into the astronaut program; I got assigned to a mission. Now I was in *space.* I didn't start crying, but if I didn't have a task to do, I probably would have. I looked out at the horizon, and I could see the sunlight passing through our atmosphere, creating the brightest, most beautiful shades of blue. It was almost fluorescent. My colleagues were going through their own checklists, so we weren't talking about our feelings or observations. But there *was* someone with me to share the emotional part of this experience—an eight-year-old kid named *Xiǎo Zhong* who played spaceman in his parents' garage. I thanked him for dreaming big, and he thanked me for doing the hard work to get us here.

Q14: WERE YOU ABLE TO GET AN EXTRAVEHICULAR ACTIVITY (EVA) MISSION?

SINCE LEONOV'S HISTORICAL FIRST, there were only ninety-five astronauts given the opportunity to walk in space when I was at Johnson Space Center waiting for my second flight assignment. Except for one Frenchman, all were either Soviet Russian, or American. I was hoping to be number ninety-six.

As a kid, I was a superfan of every one of the American spacewalkers. I was six years old when Buzz Aldrin snapped the first space selfie from a camera attached to Gemini XII during his first EVA in 1966. I was jumping out of my skin watching the Apollo 11 team step on the lunar surface in 1969. In February 1984, I was in graduate school at UC Santa Barbara when Bruce McCandless II flew through space untethered to the Space Shuttle. My jaw dropped watching the veteran spaceman maneuver a nitrogen-propelled jetpack with a couple of joysticks.

Later that year, the Soviets sent the first women on an EVA. Svetlana Savitskaya was in such good shape that they had her learn how to use heavy welding tools on Earth, then put her in a bulky spacesuit to test the tools while floating in zero-g. Just weeks after, American geologist and oceanographer Kathryn Sullivan became the first woman on team NASA to walk in space. Sullivan successfully tested the Orbital Refueling System—a way to extend the life of satellites by adding the propellant that keeps them steady while they're in orbit.

In the early days of NASA, the astronauts were all military test pilots and all white men. With the beginning of the Space Shuttle program, there was suddenly a need for scientists and engineer astronauts to perform the research experiments and other operations. That opened the door for a more diverse astronaut corps. NASA's first Asian American astronaut was Ellison Onizuka. He was part of the astronaut class of 1978, which included six women, three African Americans and Onizuka, all firsts. Sadly, Onizuka never got to walk in space because he died in the *Challenger* disaster in 1986.

By the time I was in line for my first flight in 1993, NASA's budget was flat. The public was less dazzled by astronauts floating in space as they once were, and there weren't a whole lot of reasons for missions to include EVAs. To justify the costs of sending astronauts outside during Space Shuttle missions, EVAs needed to be more than just a walk in space. The truth is, most astronauts never get a spacewalk in their entire career. It was something that filled me with dread while I still tried to remain Zen about it.

We astronauts were delighted when NASA finally got the greenlight to start construction of the International Space Station, which came out of a major redesign of the problem-filled Space Station *Freedom*. It meant a lot of work for our teams. The station would be built in a way that would need EVA astronauts to work from the outside, and of course, there would be maintenance missions over time.

Getting bumped from the Hubble repair job still lingered in my consciousness, so I was stressed about a crushing repeat. Fortunately, it hadn't left the minds of those that made the crew assignments either. They still felt like I got screwed, so they put me on the next EVA mission to make it up to me. The mission was to develop techniques and test tools that we would use to build the space station. When I got the news, I wanted to jump for joy. While the Hubble mission would have been super cool, the International Space Station would have a much longer legacy. I would be the first Asian American and the first of Chinese descent to walk in space. In a way, I felt like I was building on Ellison Onizuka's first.

Training for the EVA had us back in the pool figuring out best practices to test NASA's version of a DeWalt power tool. We joked that everything on the Space Shuttle was run on AA batteries and put together with seven-sixteenths-inch bolts. It was true for the most part, because when you're out doing a spacewalk, the last thing you'd want to do is fumble for a different size socket.

We always go on spacewalks in teams of two for many reasons, but especially in case something goes wrong. I was assigned as EV1, the lead spacewalker. I had more training, and I was on my second mission. Dan Barry, who I was paired up with, hadn't flown at all, but that didn't stop him from believing and expressing that he should be the lead spacewalker on our mission, and not me. When I found out, I couldn't believe it.

In our first training in the pool, Dan wanted the robotic arm to take him from one spot and move him to the next spot to do his work, like it would be done in space. But there were mechanical limitations. The training version of the arm was not very strong, and it was difficult for the operator, fighting gravity, to get us to the right place with the weight of an astronaut attached to it.

I didn't know how to handle Dan trying to take over.

"I'm trying, but this arm is not very strong," the operator told us over the comms system.

I was just hanging out listening to this, letting the time slip away. We got far less than half our tasks done during those first few runs.

Mark Lee was not impressed. "This is not right. You guys are not doing this right," he told us.

His deputy, Rich Clifford, stated it more bluntly: "Sounds like a leadership problem." He smiled, letting me know that he was giving me friendly advice to take charge and be firm.

I'd always excelled in the EVA training. I wasn't used to negative feedback, and I didn't like it. Rich was right. I needed to take charge.

"Now, here's what we're going to do," I said to Dan. "You're going to get out of the arm between worksites. They're going to position it where it needs to be. Then you're going to get back in and do the task. We're not going to waste time trying to have it driven from here to there."

Dan got out of the arm and moved to the next site.

Using my tactics, we quickly got to where we could complete all our tasks comfortably within the six-and-a-half-hour window. With this efficiency, we had the right amount of time margin. We were ready to fly.

Q: HOW DID IT FEEL BEING OUT IN SPACE?

THE MORNING OF OUR EVA, I felt the same fear that I had on my first launch. I wasn't worried about my life or my safety or my mom back home. I was stressed that something would happen, like a spacesuit problem, that would cancel our EVA.

Spacewalking is not like in the movies where you pull on the suit in thirty seconds, saunter into a huge airlock, push a few buttons, and go outside where it's easy to move around. In fictional accounts of EVA, the astronaut can hear their own breathing, their heart beating. None of that's true.

I gathered the parts of my suit: the stainless-steel upper torso with the backpack, the pants, boots, gloves, and the sizing rings that attach everything together. I dressed in long underwear then put on the liquid cooling garment. It's a skintight suit with cooling tubes sewn in that circulate water to keep you cool. I pulled on the pants, climbed into the upper torso, pushed my arms into the sleeves, and popped my head out through the top.

One by one, I connected my umbilicals, the water tube for the cooling garment, the communications cap to the cable in the neck ring, and the biomedical harness to the electrical connector at the waist. This was so that the flight surgeon could monitor my heart rate and EKG. The other astronauts on the Shuttle assisted Dan and me getting ready.

For EVA, we have another member of our crew who serves as the Intravehicular Activity (IVA) astronaut. The IVA read through the checklist step by step while Dan and I executed the commands and confirmed. IVA was with us during prep, the actual EVA, and post-EVA to make sure we did everything in order.

I put on my gloves and then the helmet. It felt so real now. The inner hatch was sealed. We started depressurizing the airlock mounted aft of the Shuttle crew compartment. I was calm, but also excited. I'd trained for this moment dozens of times, but this felt so different. Dan and I switched to battery power and disconnected our umbilicals. I was now a fully functioning self-contained human inside the suit. It was time to open the outer hatch.

"Okay, I'm opening the hatch and I'm breaking the seal," I called to the IVA.

I rotated the handle to retract the latches. I pulled hard on the hatch, but it would not immediately yield. I knew that the sticky rubber gaskets sealing the hatch would take a bit of force to break loose. But it was surprisingly hard.

My communication with IVA (and with all of mission control listening in) was all business, but inside, I felt an adrenaline rush.

"Okay, hatch is open, I'm starting my egress . . ."

Seeing the Earth from space for the first time on my prior mission was more emotional than I expected. After having dreamed about it for so long and worked so hard and jumped through all the hoops . . . It was just, *Wow.* But now, the view looking out from my helmet visor was very different. I'd seen videos and heard from other astronauts what it was like, but I was completely awestruck. I knew that the Earth was round, obviously. But for the first time, I had the visual impression that Earth was this gigantic sphere that filled my peripheral vision. It made the perspective even bigger. The flight surgeon told me later my heart rate spiked up to about 135 beats per minute before it settled down.

We had a long to-do list, so I couldn't spend too much time gawking at our beautiful planet and the endlessness around me.

Working in zero-g was so different than in the pool where the viscosity of water holds you where you need to be. I touched a handrail with a finger and the force sent me in the opposite direction. I reached out and grabbed the handrail and twisted 180 degrees away. I learned quickly to use two fingers and lightly touch the handrails to control myself. I started to sweat on the top half of my body (the part doing work) that was in the sunlight. My legs, doing no work and basically just dangling in space, were freezing in the shade. It was such a dramatic temperature difference.

EVA is a very physical experience. You must be in good shape, and you learn little tricks that help you work smart. I learned to hold the power tool in my left hand when I was unscrewing bolts, so that the handle would react against my left palm. That way, I wouldn't have to fight the force with my right fingers. I switched to my right hand when I was tightening bolts clockwise for the same reason. We were trying to manipulate heavy electric cables and make electrical connections that we would need on the space station, but with the temperature around minus 250°F in the shade, the cables got so stiff we could barely make them bend. We made a note that the cables would have to have a lot of slack for future teams to be able to wrangle them into place on the ISS.

We also tested a new tool called the Body Restraint Tether (BRT). It was the idea of a few EVA engineers and championed by astronaut Mike Gernhardt who hadn't been to space yet but had hundreds of hours working as a deep-sea diver on oil rigs and other projects in his previous career. It was a tether made of metal balls with a steel cable through the middle. The cable could be tightened by twisting a collar, and the metal balls would become semi-rigid. One end attached rigidly to your spacesuit and on the other was a jaw that you would attach to a handrail. This was a clever way to hold you in position and use both hands for doing work. It worked like a charm and was way less bulky than the other foot restraints that we had to carry with us on prior EVAs when we wanted to work with both hands. For some tasks that involved heavy loads, we needed the foot restraints. But, for most tasks, the BRT was compact, efficient, and easy to use.

Working in space against the stiffness of the suit is taxing. You don't want to wear out your fingers and arms too quickly. EVA astronauts work in a coordinated way. So, when there were a few minutes of free time for me while I waited for Dan to complete a task, I could rest my muscles and take it all in.

Wow. I'm really out here. I'm really doing this.

I looked back through the shuttle window at my friends and colleagues. They waved to me, and I waved back. It was odd to think that they were only a few feet away, but if something went critically wrong for Dan or me or both of us, they couldn't do anything to help. Of course, we were in constant communication with IVA (always with mission control listening in and injecting comments/commands as needed), so we had plenty of help. But, in an emergency, it was up to us on the outside.

I took a long look at the Earth and felt a wave of emotion. Despite the fact that there were people suffering and dying on our planet, the colors were so brilliant and full of life. The Earth looked so peaceful. It was hard to reconcile in my mind.

I came home with a new perspective and a greater appreciation for my own life. Before, I was the kind of person that if I felt like I got ripped off, I'd be irritated the rest of the day. If my university football team lost, I'd be upset about it for hours. I came back from space, and I learned to let the small stuff go. I'd found my Zen, a gift not from my Chinese ancestors, but from space.

Q15: WHAT WAS THE RUSSIAN SPACE PROGRAM LIKE?

AFTER MY SECOND MISSION, I was told to start learning Russian and get the travel department to book me on a trip to Moscow.

We flew Air France through Paris and landed in Sheremetyevo International Airport on a fall day in 1996. I handed my American passport to a grumpy middle-aged blonde behind the customs counter. Guards with automatic weapons flanked the booths, staring through us foreigners without making eye contact.

"Hello," I said, trying to look friendly and unintimidating. It wasn't hard. My mouth automatically formed a smile. I was sure I looked like a typical American with a stupid grin on my face.

The woman said nothing, opened my passport, stared at me with suspicious boredom, and walked away. I was a somewhat unwanted guest of their government—funded by one of the biggest budgets their space program had ever seen.

Finally, the agent returned, stamped her ink pad hard and then my passport: 15 OCT 1996: *Россия*, Russia.

Our Russian driver and main interpreter, Andrey, took me and four NASA engineers to a Dodge passenger van, a loaner to NASA from the American embassy fleet. It stood out among faded tin cans on wheels and upscale black sedans carrying proto-oligarchs driven by armed bodyguards.

I peered out the window as we headed to the capital. We passed limestone and wooden buildings streaked and worn by soot and acid rain. NASA established offices in the US Embassy in 1993 for Americans to come work with the ex-communist cosmonauts, technical engineers, and specialists. I knew a few words in Russian from a mandatory two-week class back at the Johnson Space Center, so I could read some of the Cyrillic. A grimy door had a makeshift sign: *Бар*, Bar. And on another dingy exterior wall, a hand-painted: *Ресторан*, Restaurant. Former Soviet citizens dipping their toes into business ventures had yet to be introduced to branding.

On the streets, I didn't stand out too much with my Asian looks. A large part of the Soviet Union encompassed Siberia and the border regions with China and Mongolia. It was not uncommon to see people with Asiatic features. My colleagues and I *did* stand out because of the way we dressed: colorful polo shirts with stripes on top, khakis or jeans on the bottom. Sneakers. Always sneakers. Everyone everywhere in the world can spot an American.

I went out for dinner with a NASA friend and flight operations specialist who'd already spent a lot of time in Moscow, and I looked forward to my first authentic Russian meal. I was shocked when we arrived at the restaurant. The familiar sign, so big, bold, and a comically bright yellow in the sea of gray.

"Why are we eating *here*?" I asked my buddy John McBrine.

Leaving the streets full of stone-faced pedestrians into an American fast-food restaurant was a shock. Young women at cash registers greeted us with giant smiles. Although I didn't yet know much Russian, it was clear what they were saying. "Welcome to McDonalds. Can I take your order?" Just like America.

Several of my colleagues had been to Moscow before.

"This is our go-to restaurant here in Moscow," John said. "Because when we eat at McDonald's, we know what we're getting." He explained that with food and electricity shortages, you never knew how long meat was sitting around in a local restaurant. Most everyone from NASA who ate locally got sick at least once from rotten food.

"McDonald's has its own sourcing and American-style quality control," he told me, shrugging. It was a microcosm of America in the middle of Moscow.

The rest of the guests were elite Russian businessmen with their families and black-market entrepreneurs with fistfuls of American dollars. A Big Mac, apple pie, and a milkshake would set the average Russian worker back two days' salary.

We checked in to the Penta hotel, one of the only upscale "Western" hotels in the city. It was built by the German airline Lufthansa for the 1980 Olympics—the games we boycotted after the Soviets invaded Afghanistan.

I slept fitfully to blasting night club beats and occasional explosions, likely of some kind of conflicts between rival mafia gangs, which seemed to rule Moscow at the time.

After spending a few days touring the space facilities in Moscow, we headed to the functional equivalent of NASA's

Johnson Space Center. It was the epicenter of the Russian space program, named after the greatest space celebrity of the east, Yuri Gagarin, the first human to fly into space back in April 1961. Yuri Gagarin Cosmonaut Training Center was a walled, formerly secret military base, informally known to everyone as Star City.

Our driver piloted our van along a narrow two-lane road heading northeast. Chaotic Moscow gave way to golden birch trees shedding their leaves before winter.

"Just four years ago, Star City wasn't even on the map," Andrey said. "It was top secret."

I'd just finished my second shuttle mission for NASA. I'd been fortunate to do two spacewalks, which made me an Extra Vehicular Activity specialist. Now I was collaborating with Roscosmos on building the International Space Station. What type of tools and construction techniques did they use on their *Mir* space station? How did they pilot their reusable Soyuz capsule? How do their spacesuits compare to ours? How do they prepare for and perform EVA? I needed to learn all of it. This was part of the program that NASA termed *Phase One*.

I had done my homework. All cosmonauts train in Star City. Their families lived here. Their kids went to school here. Some grandkids of their first space explorers still lived here. It was a self-contained city. Andrey estimated the population to be about six thousand behind the walls.

Unlike Johnson Space Center and NASA, Star City, as well as the entire Soviet space program, was an arm of the military. Everything was completely clandestine during communist rule. Only in the mid '90s was the space program connected to the civilian-government space agency called Roscosmos.

I peered out at the landscape. Along the narrow road,

houses, buildings, and small shacks were painted the same drab olive green.

"Why is everything here the same color?" I asked Andrey.

"One day during the Soviet times," he told me, "a VIP was coming to visit Star City. The commanding officer of Star City only had a day's notice. He wanted things to look nice, so he ordered his soldiers to go work all night and paint everything along the road from Moscow to Star City. Of course, they only had army green paint."

I wasn't sure it was a true story, but I couldn't think of a better explanation.

We approached the big metal gates of Yuri Gagarin Cosmonaut Training Center. Uniformed soldiers checked our papers and waved us onto the campus. I looked around at the weathered buildings, Soviet architecture, and magical taiga forest, and felt like I was stepping back in time.

Star City was a wild place in so many ways. The whole compound felt like a museum. Offices and laboratories were monolithic concrete buildings. Inside, run-down rooms had ornate wool carpets handmade by Caucus and Central Asian artisans. I met some of the Soviet space legends who were retired and still living in Star City, or Moscow. There was Dr. Abramov, the leader of the team that created their first EVA spacesuit in 1965. There were engineers that worked on their Moon program in the 1960s when they were competing with our Apollo program.

The cultural differences were stark.

For meals, we lined up in the dingy cafeteria with plastic trays and aluminum flatware. It felt like a prison. Behind the counter, an old Russian guy in a filthy apron would spoon a piece of meat from a giant tray and press it onto my plate with his thumb. He did that for everyone, every single time.

NASA bought some small, prefabricated houses for our teams that stood out as a tiny American village against the blocky soviet architecture. We put on athletic gear and went through an opening in the wall to a trail in the forest—to run. For exercise. It was bizarre to the Russian team. Everything about how we lived and operated seemed to miff them.

"Why are you Americans always running? *What for?* And why are you always carrying water bottles?"

Their pastime outside of the compound walls was to build small *dachas*—shelters where they would tend a small garden, smoke, look for mushrooms, drink vodka, and hang out with their buddies. That's what a forest was for.

They also thought we looked weird.

"Why do you people all dress like children?" one Russian secretary asked me.

"What do you mean?" I asked.

"It's just . . . you all wear these colorful clothes, and you have these puffy jackets that have color on them. You look like children. It's ridiculous."

Mostly, we took their insults in stride. But if they talked trash about our space program, it got really annoying. As annoying as they were, there was a more endearing side to cosmonaut cockiness that struck me. Their seventy-year-old government and its funding had vanished from under their feet. There was a real question on everyone's mind whether Russia would survive as a cohesive country. Store shelves were empty, the cities were in chaos, and the government couldn't pay for anything.

And yet every morning there was a beautiful fresh bouquet of flowers on the giant statue of Yuri Gagarin just inside of the main gate. That's how much they all loved their space program and respected the cosmonauts and everyone who made their many successes possible. Standard operating procedure for Roscosmos' technical program was a completely different ball game than at NASA. Their Mission Control Center on the outskirts of Moscow—the place where they ran the space missions like we did from Mission Control in Houston—had a cafeteria where you could get some snacks and tea. It also had bottles of liquor on a shelf. The first time I went there, two guys were sitting, chatting, and drinking vodka. It was nine in the morning.

"Wait a minute. These guys, they're drinking *alcohol*?" I asked Andrey. "It's 9:00 a.m."

"They probably just came off night shift," he said, unbothered.

The difference in our equipment was night and day. At NASA we had Cadillacs of space machines that we spent millions on optimizing for performance. But because they were computer controlled and complicated, there were a lot of things that could break and go wrong—extremely capable vehicles and equipment with complex failure modes.

The Russian engineers built space gear like an old pickup truck—AM radio, no air conditioning, hand-roll-down windows. The technology was not sophisticated, but it was reliable, and it worked.

I trained in the Soyuz flight simulator to get a feel for what it was like flying in their spacecraft. I clenched my teeth in their centrifuge, where they spun me up to eight G's until my eyeballs flickered uncontrollably and I couldn't see a thing. (We only get up to five G's from a real rocket launch, so I wasn't quite sure what the point was.)

The most important thing I needed to learn was how they did spacewalks, since that was my specialty for the ISS program. There was a huge round building that contained a giant water tank and a full-size mockup of *Mir* space station that they lowered into the water to practice working in space.

I got dressed in their long underwear then put on the liquid cooling garment—an adult onesie with tubes for water to keep your skin cool while working in the sweltering spacesuits. I pulled on the communications cap and fastened the chin strap. Then I got into their suit. The Russian engineers were all very proud of their spacesuit.

"You realize, we created a spacesuit before you guys did," one of the technicians said to me as I opened the unlatched back door of the suit to swing my legs in. "And you can see how very excellent it is."

"Of course," I replied with my American politeness. "It was a great achievement."

Their suit wasn't functionally much different than ours, but there was no custom fit. The fabric in the arms and legs scrunched up after the cables were pulled tight to fit me. I fastened the water line to my cooling garment and the communications connecter to my headset so I could hear them, and they could hear me.

Unlike at NASA, where EVA astronauts had customized gloves to make it easier to work in space, they had medium and large. I took the medium ones and snapped them into the suit. There was only one small hinge to lock them into place—whereas my NASA gloves had three very deliberate steps to attach and release. *Dang. What if I snag this on something*? I thought. *It's like, Oh my God, I'm dead.*

"You don't have another lock over that?" I asked one of the specialists.

"No. Why would we need it?" He looked at me like I was the biggest wimp ever. Apparently, the idea of extra safety features was unnecessary for cosmonauts.

They put me in a vacuum chamber and pulled all the air out so that I could evaluate their spacesuit under conditions similar to space. The building with the vacuum chamber was rundown and dirty. Even more unsettling, the technicians—the men who would have my life in their hands—seemed unengaged. They were sitting in chairs around what looked like a coffee can using kabob sticks to roast tiny fish over a fire.

"Wait," I said to our interpreter Andrey, "are these the guys that are running the vacuum chamber? I'm going into this vacuum run, and these guys are sitting around eating what looks like goldfish from the lobby aquarium?"

"Don't worry. Not a problem," Andrey said.

I survived the vacuum chamber. Next up was the underwater run in their hydrolab. They pressurized my suit to how it would feel in space. The suit inflated and stiffened. Every time I moved, I was working against the pressure inside of the suit. It also stretched out the arm and leg segments, so my suit no longer fit me well.

Two cranes lifted me and my Russian instructor up above the tank, swung us over the water, and lowered us in. Divers disconnected us from the cranes and put weights in our suits until we were neutrally buoyant, just like we did at NASA. We were ready to simulate a spacewalk. But with the awkward fitting Russian suit, I felt like I was floating inside a balloon. I struggled with the tasks. I worked hard to connect and disconnect electrical cables and fluid lines. I was breathing hard, turning bolts, and installing equipment. It wasn't easy, but I got everything done and surfaced about five hours later.

Then I got subjected to a Russian debrief.

The instrumentation system recorded my vital statistics. "Your heart rate was much too high. You were working too hard. That's not the way to do it. You Americans obviously don't know anything about doing EVA."

Yeah, okay, whatever. I'm just here to evaluate your equipment. As rocky as our first partnerships were in those early days, I grew to trust and respect these guys like my own NASA colleagues. Over the next three decades, wars in the Balkans, Caucasus, Crimea, and Ukraine reinstated Russia as our biggest global adversary—and America as theirs. Throughout the chaos, the spacemen and spacewomen of NASA and Roscosmos held on tight as our countries' volatile relationship threatened to leave us conflicted in space and cripple our future collaborations.

Q16: HOW WAS IT FLYING WITH THE RUSSIANS?

IN LATE 2000, I was assigned as the backup commander of Expedition 8 to the International Space Station, which was a collaboration with our Roscosmos partners. When you're first assigned to fly to the ISS for a long-duration flight, you start training as part of a backup crew. You train alongside the "prime" crew—the group set to go up before you. Should something happen to the prime crew, if somebody gets hurt or sick or has a family emergency, then someone from the backup crew—or the entire backup crew—is ready to go. It happens more often than you might think.

It was with mixed feelings that I started my training in late 2001. I knew that learning a new language and culture and splitting my time between Russia and Houston would change me. I knew my perspectives and attitudes would be altered. I knew that my relationships would be different too.

It was a big commitment to make, and I knew it would make my life different from then on. I was excited, but mostly curious about who I would become.

For the next several years, my fellow Americans and I would train side by side with Russian cosmonauts on a myriad of systems, procedures, experiments, and spacesuits—in both countries and both languages. We would also travel to the Canadian Space Agency headquarters in Longueuil, Quebec, to learn how to use the robotic arm they developed for the ISS.

The plan was for us to ride up in a Space Shuttle and spend several months at the space station. Shuttle flights were not without risk, but NASA felt that we had our arms around the issues and dangers. It had been seventeen years of successful missions since the *Challenger* disaster.

In Star City, we finished training for the day on February 1, 2003, Chinese New Year. I was cooking a big batch of stir-fry for our crew when the telephone rang. It was my good friend and astronaut classmate Bill McArthur.

"Turn on CNN!" he cried.

"What happened?"

"Just turn on CNN!"

Oh my God.

Over and over, the network replayed video of the final moments of Space Shuttle *Columbia*—carrying seven of my friends and colleagues—as it entered the atmosphere. A small area of the left wing had been damaged during launch. Upon re-entry at 17,500 miles per hour, the hot plasma generated by frictional heating of the air against the ship burned the wing from the inside out. Inside the vehicle, the crew realized they had a problem. Post-accident data analysis showed that commander Rick Husband struggled to control the ship in its final moments. There was nothing he could do. Seconds later, the vehicle broke up above North America and rained down to Earth. My friend's bodies scattered with the debris across northern and central Texas.

I was in shock. I felt sick, depressed, lost. People I saw in the halls of NASA every day were dead—my friends who had the same passion for space that I had. These were men and women with families, children, and friends who had just watched them launch into space two weeks before. Now they would never come home. The celebration dinner I prepared became a somber remembrance for our fallen friends. We raised our glasses to them in silence.

NASA grounded the Space Shuttle fleet. Americans on the ISS had no way to get back to Earth, and everything was uncertain. I believe that one thing all humans dislike is uncertainty, and that was all we had during that time.

"Do you think this is the end of the Shuttle program?" asked one of my colleagues.

"I don't know," someone else replied. "I'm not sure what's going to happen."

We stopped training. We didn't travel to Russia, and the cosmonauts didn't come to train with us. Everything about our future was in limbo.

Fortunately, the paralysis didn't last. As with other catastrophic events that claimed the lives of astronauts, NASA worked out a plan to get us back into space. This time, Americans couldn't go alone, and fortunately, our former adversaries were now long-standing space partners. Just as America and NASA had come to Russia's rescue during the chaos at the end of the Soviet Union—flying several critical Space Shuttle missions to their *Mir* space station—Russia would fly NASA astronauts to the ISS on their Soyuz spacecraft until the Space Shuttle could fly again.

The missions had to change significantly. Soyuz is a much different vehicle than Space Shuttle. For starters, it's tiny. It fits just three people compared to the Shuttle's seven. Also, Russia could only produce so many Soyuz and Progress resupply vehicles, given their meager budget and aging infrastructure. We were originally scheduled to be on board the ISS for four months, but these limitations stretched the missions to six months. Still, Russia was the only game in town. I was grateful for their help as I headed back to Star City for training.

Q: WHAT WAS STAR CITY LIKE THEN?

I'D BEEN BACK TO Russia a few times since my first visit in 1996. In 1997, Russians collided a cargo ship into their *Mir* space station. My NASA colleague, Michael Foale, was near the Spektr module that was hit and leaking air. The vacuum of space threatened to suck all the air out of *Mir*. Foale and his cosmonaut colleagues felt their ears pop from the loss of pressure but were able to close the hatches and seal Spektr off. They saved themselves and the station.

The *Mir*, originally Russia's crown jewel of its space program and the only space station to be permanently inhabited since it launched in 1986, had become a bucket of bolts. It was designed for a six-year operational life but was pushing twelve. In the words of one senior NASA Flight Director, "The wheels are falling off." *Mir* had several near catastrophes including a serious onboard fire.

After the breakup of the Soviet Union, newly independent Ukraine wanted to be paid for their "Kurs" auto-docking system that was used by all Russian spacecraft flying to space stations. The Russian space agency had to pay, but also decided to develop their own manual docking system they called "Toru." On the first test of Toru, the Progress resupply vehicle crashed into *Mir*.

Many people at NASA were furious. Why were we working with these clumsy yahoos to build the International Space Station when they didn't even have the funds to pay for their share? The American media was not kind. *The Washington Post*, somewhat unfairly, described *Mir* as "a complex array of trailer-like modules that have been attached to the *Mir* core like tinker toys over the years . . ."

Since we had an astronaut aboard *Mir*, NASA wanted eyes on the Russian space agency and their repair plans. There would undoubtedly be several EVAs to fix things from the outside, so I was sent as an experienced EVA astronaut to look over their shoulders and offer my expertise. It was far from comfortable. Things were even more adversarial than the first time I was there. I didn't start serious Russian language training until 2001, so I still needed our translator Andrey's help.

Why are you here? It's OUR space station. We know more than you anyway, Andrey kindly relayed to me.

They clearly didn't consider me a colleague. My NASA engineer team pushed the Russian team to put me into their water tank in a spacesuit and test the procedures they were developing.

"*Nyet*," was the firm reply, followed by "*Za chem?*" "What for?"

Remembering my first Roscosmos tank experience the year before when all they did was criticize my work, I told my NASA engineer colleagues that I could see enough by observing them working in the tank. There was no need to put me in the water tank.

Fortunately, *Mir* got fixed, and Foale made it back safely with his cosmonaut crewmates. After the incident, NASA astronauts were skeptical of flying aboard *Mir.* One even backed out of a mission. *Mir* hobbled along for four more years.

A couple of years after the *Mir* collision, Congress conceded to putting up the funds for the International Space Station. With such an enormous technical and financial commitment, we knew we would be partners with the Roscosmos long into the future. We were all glad that we would have a lot of future work in space. But our NASA teams were still iffy about a long-term relationship with our former foes.

I started traveling to Russia half-time in late 2001 to train for my ISS mission that would happen around 2004. Moscow had changed quite a bit in the years since the collapse of the Soviet Union. There were many more McDonald's, and now a bunch more Western fast-food joints like KFC and Subway. Women wore fitted clothes in bright colors. Skyscrapers rose up overnight. Instead of gray, monolithic structures, new buildings had Western flair and color.

The United States paid for the lion's share of the ISS program, including large payments to Roscosmos. These payments went toward many things, including construction of the Russian Zarya ISS module, astronaut training and launch, and even Russian specialist travel expenses to Houston for meetings and workshops. After a while, we got to know each other very well—not just the astronauts and cosmonauts, but the engineers, mission control specialists, and all the support folks too.

By now, my Russian was better, and I strove to speak to everyone in Star City in their language. I only spoke English to my American colleagues or when I called folks back home.

Roscosmos was very much more male dominated than NASA in the early 2000s. Over the course of the Russian space agency's history, they only averaged one woman in space per *decade*. For much of the time I was training in Star City, there was only one female cosmonaut in training. Management didn't want her there. Some older cosmonauts grumbled about having a woman in the corps. She eventually left when it became clear that she was never going to have an opportunity to fly.

With the end of the Soviet Union and its false facade of equality and egalitarian rhetoric, the nation's attitudes were laid bare. Women had fewer opportunities in most professions than men. In Star City, secretaries were female, instructors were men. Practically all the engineers in the Mission Control Center were men. Russia still seemed like 1950s America. Culturally, things were like the television show *Leave It to Beaver.* When we went to someone's house, the women served the men. It was weird for me, having grown up with parents who raised us to treat men and women as equals at home and in their careers.

Because I would be Flight Engineer One (FE-1), or copilot, on the Soyuz spacecraft, my training was going to be more intense than it would have been on the Space Shuttle. The three-seat Soyuz has a commander in the middle. On the left is Flight Engineer One. The right seat is Flight Engineer Two, who is essentially a passenger on the flight, but an important crewmember once in space.

I trained on all their spacecraft systems and spent hundreds of hours in simulators practicing nominal operations, malfunction procedures, and emergency procedures. I worked in sync with Salizhan Sharipov, my cosmonaut crewmate. We'd spend six months in space together. He would command the Soyuz on the way up. I would command the International Space Station once we got there. During most of these training sessions, it was only the two of us. It was challenging and interesting and a lot more work.

Salizhan was the first cosmonaut from Kirgizia (no one from there calls it Kyrgyzstan), and his family was originally from Uzbekistan. He looked like an Uzbek version of me. We both have spiky thick hair parted on the left. We're about the same build and roughly the same age and have a similar temperament. We spent a lot of time together. Besides training, we hung out in the Star City pub and in the on-site sauna. Sometimes we traveled to a neighboring town to a nightclub of sorts. I got to know his wife and two young kids; a daughter, Nigara, and son, Zhakhongir. Having grown up with two sisters, I called him the brother I never had.

A few days before our launch, we were invited to visit Alexei Leonov, the first man to walk in space. Just months into his seventieth year, he still lived in Star City in a "cosmonaut dacha" with a nice garden in the back. Everyone knew the wild story of how he nearly died trying to get back into the inflatable airlock of his spacecraft. With four spacewalks under my belt (I'd completed two more on my third Shuttle mission in 2000), EVA was my specialty. Here I was, in the presence of the man who did it first, all those years ago. I was the astronaut equivalent of starstruck.

His hairline had receded, and he'd put on a few pounds, but Leonov still had that cosmonaut look. He was jovial and friendly as he poured shots of vodka and served us pickles he made from the cucumbers in his garden—*zakuski*, snacks to go with our drink.

"It's quite something, what you have accomplished with the ISS," he said to me.

The ISS was about halfway finished, and I was very proud of my part in it and grateful for Leonov's stamp of approval.

"Thank you for the compliment," I said. "It's been a great collaboration. Your garden and these pickles are quite an accomplishment as well. They're delicious!"

We all lifted our glasses.

"You are about to go off on quite an adventure," Leonov said, smiling widely and clinking his glass to all of ours. "To your health!"

Being blessed by this space legend before embarking on our flight? *Ten out of ten.*

Leonov later gave Sergei Krikalev some pickles to take on his next flight—which happened to be to take over the space station after mine and Salizhan's mission. It was such a nice surprise when Sergei brought out Leonov's homemade pickles to share with us as a special treat the night they arrived. In 1955, the barren landscape of the Kazakh Steppe was where the Soviets chose to test their first rockets and intercontinental ballistic missiles. In the middle of nowhere in southern Kazakhstan, the Baikonur Cosmodrome became the site for many Soviet and Russian rocket launches and space missions. After Yuri Gagarin launched on his historic first human spaceflight from Baikonur in 1961, every USSR and Roscosmos human spaceflight since has launched out of here.

Two weeks before our launch date to the ISS, we arrived in Baikonur. There were sleeping quarters for us—aging dorm rooms that hadn't been updated in decades. It's maybe the most elite one-star hotel in the world. We had a common area with a projector and screen for nightly Russian movies, a large billiards table, a gym with some primitive workout machines, and a small computer room where we could check email. As with every building in Russia and the former Soviet states, there were ornate Central Asian wool carpets on the parquet floors.

We spent our time studying and going for runs on the steppe along the river. An occasional camel crossed our path. An old-style locomotive might steam by. Compared to Russia, Kazakhstan was even more of a throwback.

One day, I was out for a run with my American colleagues from our backup crew when we all came to an abrupt stop. On the crude dirt path, just ten meters in front of us, was a land mine. We had no way of knowing if it was live or a dummy training one, but we weren't about to find out. We turned around, ran back, and told our Russian handlers about it. We never did learn why it was there or if it was live.

Q: HOW DID THE SOYUZ COMPARE TO AMERICAN SPACECRAFT?

IN ORDER TO GET into the same orbital plane as the ISS, we had to be very precise when we launched the Soyuz or the Space Shuttle. Timing is critical. The week before launch, we sleep-shifted to change our body clocks, going to bed a bit later each night.

We woke up a little after midnight on October 14, 2004. I showered and put on my blue flight suit adorned with my name badge and our Expedition-10 mission patch. Our patch had a very cool design: A US flag crossing the Russian flag to form the letter X—Roman numeral 10. I was handed a black Sharpie, and I signed the wooden door of the room, another tradition at Baikonur. I proudly wrote my name in English, Russian, and Chinese for good measure next to dozens of historic names of cosmonauts who stayed in the same room before me.

Salizhan, Yuri Shargin (Flight Engineer Two), me, and our backup crew went to the canteen for our prelaunch meal. Then we met up with the head of the cosmonaut corps, Valery Korzun, along with members of the Russian leadership from Star City and Baikonur. Korzun popped the cork on a bottle of sickly-sweet Russian "champagne" and poured a glass for each of us.

"Oh, okay. So, we're drinking on launch morning," I laughed with raised eyebrows. It was so *not* NASA.

"*Normalno!*" Salizhan said. "Normal. It's okay!"

I knew that we would not drink to good luck. Wishing someone good luck in Russia is what you do to jinx them. We toasted to a successful journey.

We drove across the dark landscape to the building where we put on our pressure suits. It was a cool, starry night. Perfect.

Russians are intent on their traditions and superstitions. Before you embark on a long journey, you all do a "*za poseshok*"—drink a shot of something, usually vodka or cognac. Then you sit on your suitcases or crouch down and reflect for a few seconds. Then whoever's the leader stands up and says, "Let's go."

Yuri Semenov, the chief designer of the Russian space station program—the same guy that started the whole thing back in the 1970s, sat at the table in the small room of the suit-up building. He invited us to sit at a small table with glasses filled with cognac.

"You are embarking on an important mission," Semenov said warmly. "The future of the program depends on each mission's success. Just as this was true with *Salyut*, then *Mir*, now it is true with the ISS."

Semenov picked up his glass, downed the cognac, and we followed. We all paused for a few seconds. Then he said, "Time to go!"

Salizhan had been chain-smoking almost the entire time I'd known him. He ran outside to have a cigarette.

"Are you going to be okay, like, with no cigarettes?" I asked him. "Because we're going to be gone for more than six months."

"It's no problem," he said.

He could tell I wasn't convinced.

"Look," he said, "we all smoke. But when we go to the station, it's not like you can go out and get a pack of cigarettes. So, you just put it out of your mind. It's not even an option. It's not a big deal."

We got into our white Sokol pressure suits. Our backup crew stayed in their blue flight suits. Then, like some sort of television makeover reveal, a curtain opened and behind a big window was a large group of reporters and family members snapping pictures of us through the glass. It was three thirty in the morning. I posed awkwardly.

We were directed to go outside, line up three abreast, and march toward the head of the commission, who would formally give us permission to board the spacecraft and fly. Salizhan, as the commander of the Soyuz, was in the center. I was on his left, and cosmonaut Yuri Shargin was on his right. We planted

our feet on white squares labeled with our positions in the spacecraft. In unison, we saluted. Even though I'd never been in the military, I decided at the last second to salute, which felt a little weird. (Salizhan and Yuri were both Russian military.)

"We have completed our training and are ready to fly," Salizhan announced.

The head of the commission, a government official I had never seen before, said, "I grant you permission."

We waved to everyone as we got on the bus to take us to the launch pad. Yuri was a rookie. Salizhan flew aboard a Space Shuttle some years prior as part of Phase-I of our collaboration with Russia. I was the veteran, having flown on three Shuttle missions and having done four EVAs. Yuri would fly up with us, spend around a week onboard during the handover period, and then go back down with the Expedition-9 crew in their Soyuz.

They didn't want us getting sick before we left, so we'd quarantined for two weeks, only interacting with a few medically cleared team members. But when the bus stopped, we got dropped into a mob of people. Senior bosses and some senior engineers were there, grabbing us for bear hugs, patting us on the back, and kissing us on the cheek. The rocket vented liquid oxygen surrounding us all in puffy clouds like an '80s music video.

So much for quarantine.

Since arriving at Baikonur, there was always a young, fit, KGB guy nearby. He stood out in his long brown coat, listening to everything we said. He never said a word to us. Today, he followed us to the launch pad. Then he walked up the stairs, put us in the elevator, and watched the door close, completely stone-faced. He didn't even say goodbye.

We got to about 150 feet up where the Soyuz spacecraft sat atop the Soyuz rocket. There was a platform around the spacecraft where we met the technicians who checked our suits and helped us into the Soyuz.

Yuri was Flight Engineer Two, so they put him in first. They put me in second on the left. It was a mild surprise. This vehicle was *tiny*. Of course, I'd been in the Soyuz simulator for hundreds of hours, and the actual vehicle before, in shirtsleeves. But the simulator is a mockup with the side door cut out. It's easy to get in and out, and they always leave the door open. But this was real life. The sardine-can metaphor was as valid as you could get, especially dressed in the bulky pressure suit. I started getting a weird feeling.

This must be a bit like what it feels to be claustrophobic. I shook it off. But still, I was like, *Damn, this is really small in here.*

Almost fully reclined on our backs (except with knees pushed toward our chests), we strapped in to our five-point harnesses. We had about two hours to go through the prelaunch procedures and checks. We connected the communication cables, the ventilation hoses that blow cool air into our suits, and the oxygen hoses—to be used in case of emergency. We performed our suit leak checks and went through the entire written checklist.

Above us was the hatch that led to the rest of the spacecraft. The upper module is an additional living volume that you can open once you're safely in space. It would allow the three of us to spread out a bit. It was also the storage area where we were bringing up our extra food to replace what the last guys ate. Everything was bolted and strapped down so it couldn't move.

We listened to what was happening on the ground and followed the launch sequence.

The countdown was now under a minute.

It's real.

My mind was completely empty of stray thoughts. I listened intently to the count as my left hand held our nominal and malfunction procedure books. My right hand was relaxed on my thigh, ready to issue commands to the computer (if necessary) as we "flew uphill."

Ten . . . Nine . . .

The engines were lit before T minus zero, and we heard the light rumble. At T minus zero, the rocket gently rose from the pad.

It was a very different experience than the Space Shuttle. At T minus zero on the Shuttle, there is a loud explosion as the solid rocket boosters ignite, adding an additional five point five million pounds of thrust to the one point five million pounds already generated by the cryogenic liquid-fuel engines on the Shuttle vehicle. You literally explode off the launch pad.

For the Soyuz, they use a rocket with only liquid engines. The thrust builds gradually, and the rocket gently rises off the ground. You don't even feel it. Except for the excitement, it was almost Zen. After arriving in orbit, we flew autonomously for about two days at an altitude of around 250 nautical miles before arriving at the right spot to begin our approach to the ISS for docking. Within two hundred meters of the station, the master alarm sounded. Warnings appeared.

What the hell?

I'd trained on all possible emergencies. In the simulator, alarms and warning lights constantly go off, and we would work on the appropriate procedures to get ourselves out of trouble. But in actual flight, things usually go pretty much as planned, without blaring alarms.

"The autopilot is speeding us toward the station," Salizhan said.

Shit. It should have been slowing us down.

"We're rotating to the left," I said. We were losing visual sight of the ISS through the periscope. We were on a collision course with the station.

Salizhan and I had our malfunction procedures out. We flipped through to the appropriate page. I went down the checklist.

"I'm enabling manual control," I told Salizhan. This would take authority from the autopilot and allow him to fly the spacecraft himself.

Salizhan righted the ship back into alignment with the station. He then fired braking burns, using the thrusters to slow our approach.

Excited chatter from the Russian Mission Control in Korolev filled our ears. It was hard to understand. They were talking over each other, telling us what to do. I realized they were behind us in the checklist.

"We have stabilized the ship manually," Salizhan calmly told them. He engaged short thruster firings to keep us in relative motion to the station. "We're station-keeping with ISS at a range of fifty meters. Permission to execute the docking manually?"

"*Nyet*," the rendezvous flight controller said. "Re-engage the automatic system."

What? Salizhan and I looked at each other, puzzled. *Why should we do that?*

Salizhan tried to change the flight controller's mind. It was futile.

Salizhan nodded to me to proceed, and I executed the commands. Instantly, the alarms blared. Thrusters started firing, moving us out of position. I could feel my heart pounding. Without being told, I quickly gave manual control back to Salizhan, and he again righted the ship using the manual hand controllers.

We exchanged a look, sweat beading on our foreheads. Without speaking, our mutually furrowed eyebrows made it understood that no matter what the ground said, we were not going back on autopilot.

Fortunately, after a short discussion, the flight controllers cleared us to dock manually. Salizhan nudged our spacecraft forward and performed a perfect approach. The soft-dock mechanism caught the extended probe of the Soyuz, and we began the procedure to retract it, pulling us into contact with the station docking port. We activated the latches, locking us into a hard connection with the ISS.

I took a deep breath.

Congratulations, man. Excellent work, my smile said to Salizhan.

Yes. That was a close call. You did good too, his return smile said to me.

We took our right gloves off and shook hands (it's bad luck and rude for a Russian to shake hands with gloves on, whether on Earth or in space).

What caused this problem? The specialists on the ground shared their findings with us weeks later. There were two failures. One of the thrusters failed to fire when commanded by the computer, then failed to inform the computer that it hadn't fired. So the computer commanded more thrusters to fire in a constant loop, which had the effect of speeding up and rotating the ship. Taking manual control allowed us to fly the vehicle through another system, which saved us. A simple software patch upload blocked the failed jet out of the set, so that we wouldn't have that problem again on our return trip.

After all those years flying Soyuz (since 1967), this was the first time this failure mode had occurred.

Once we were safe, I took a moment to breathe and think about what had just happened, and I felt the hairs rise on the back of my neck. If we had collided with the station, we would almost certainly have been lost, and such a collision could have easily caused a huge leak onboard the ISS that the crew might not have been able to survive.

Q17: WHAT WAS THE WILDEST THING THAT HAPPENED TO YOU IN SPACE?

SEVERAL MONTHS BEFORE our launch, Salizhan and I ordered food for our entire mission. Food in space works like this: We are given a long list of menu items from both the Russian agency and NASA. Since we were on a dual-country mission, we could choose food items from the NASA menu as well as the Russian one. I had some favorites that I liked from previous trips and food tastings. I marked down several cans of Russian fish with tomato sauce, mashed potatoes with onions, and *tvorog*, something akin to cheese curds with nuts, sweetened with raw honey. We also got a bonus container of food with snacks not on the main menu. Mine had Pringles, packaged cheese dip, and Japanese curry rice.

Not long before launch, we got word that the crew of two aboard the ISS, American Mike Fincke and Russian Gennady Padalka, were running low on their food rations. They got clearance to get into our food, which was shipped up ahead of us and stored on the ISS. The Russian medical team and our NASA food folks told us they would pack what Fincke and Padalka put on the list of things they ate, and we'd bring replacements with us aboard the Soyuz. No big deal.

We arrived on the ISS after our scary docking episode, got out of our pressure suits, and changed into our soft flight suits. With the crew already there, we went through the procedures to equalize the pressure between our two spacecraft. It was time to open our hatch. As usual, we had to struggle with the rubber seals that often stick to the metal, making it hard to get loose. Salizhan unlatched the hatch and tugged on it, to no avail. I grabbed the mechanism, and we pulled and pulled. On Earth, you can put your weight into it. Here in zero-g, you have to position your legs to be able to push against it. The hatch finally budged and swung open.

We got a warm welcome from the guys on board, the Russian commander Padalka and flight engineer Fincke. They were happy because our arrival meant they got to go home. We were excited about starting our mission. In the Russian tradition to welcome someone to your home or back from a trip, you share bread with salt sprinkled on it. In space, we can't have salt crystals floating around, so we used small bottles of salt water instead.

Padalka handed us some tiny, bite-sized loaves of Russian black bread and liquid salt. "*Prekhodi!* Come in!" he exclaimed.

"*Spasibo!* Thank you!" we replied, sharing hugs and smiles all around.

We spent the next ten days doing our handover. There were some awkward moments. Fincke was the type of guy who was often so effusive with compliments that it felt insincere. He'd say things like: *Hey, Leroy, you're the BEST!*

Man, I wish I was as good as you!—with what seemed to me like a disingenuous smile. It made me question his motives. It certainly wasn't just me, by any means. His reputation was widely shared in the office. He was generally a nice guy and very good technically and operationally. But I thought he could also be a bit of a suck-up around superiors on both sides of the Atlantic. Padalka was one of Roscosmos' senior Air Force golden boys who had logged a lot of time in space. He was quite competent but sometimes came across to me as a know-it-all.

They gave us a tour of the station, showed us where the fire extinguishers and emergency breathing masks were. We went through the different places where trash bags and wipes were stowed. It was wild to be here. I'd been inside when I helped build the ISS on my third mission, but it was so much bigger now.

After ten days, we put Padalka and Fincke in their Soyuz. They undocked from the ISS and flew home. I waited for them to come around the Earth and took a photo of their spacecraft as it streaked underneath us, entering the atmosphere like a fiery comet.

For the next month, everything seemed fine. There are a lot of essential and moving parts on the ISS that need to be maintained. With a crew of only two (Yuri went down with Fincke and Padalka), we were busy most of the time. We changed filters and replaced fluorescent light assemblies. It could take a full day to remove an old pump and install a new one. We organized and cataloged supplies and equipment and gathered the items destined to be thrown away when the resupply ship came in (the trash goes out in it when it leaves). A lot of the time, we were the cleaning crew. *So glamorous.*

We mostly worked alone on our own schedules sent up from our respective mission control centers. We got together at dinnertime since our lunch periods were not necessarily aligned every day. We had our meals at the main table in the Zvezda module, the heart of the station. It may seem odd to have a table in space. With no gravity, what's the point? But the Russians always had a table in their stations' main modules, because humans in every culture gather around a table to share meals. There was something oddly comforting to do this familiar thing every evening.

In Zvezda we had a sound system. It was just a car CD stereo unit with speakers, mounted in crude, custom-made wooden boxes. I was never into country music, but Salizhan became a big fan after spending so much time in Houston. He brought lots of country CDs and played them during dinner. I couldn't help becoming a fan of the songs about loving pickup trucks and freedom.

We already knew each other quite well from training and socializing together for years, but dinnertime was when the two of us had only each other to talk to. We told each other about our childhoods and upbringings. He said that around the same time I was in second grade dreaming of being a test pilot in America, he was a kid in Kirgizia with a similar dream.

"One day, I looked up in the air and saw jet fighters flying," he said. "It was the beginning of my path to become a cosmonaut." At the time, his father was a shepherd. Not only did he not have a concept of what it meant to fly into space, he didn't even fully understand what it meant to be a fighter pilot.

Aside from cleaning and fixing things, we did have some cool experiments to conduct. My favorite was a remote ultrasound technique developed for spaceflight. We were trained on what to look for in the images. When you first look at an ultrasound image, it looks like a bunch of splotches. But with a little experience, the human brain starts to figure out what's what. We had a surgeon named Dr. Scott Dulchavsky and his team on Earth guiding us from the NASA Mission Control Center in Houston. Dulchavsky and Dr. Ashot Sargsyan looked at our real-time images and guided us on how to adjust the probe to make them better.

I held the wand to Salizhan's closed eye.

"Move the probe up and rotate it slightly clockwise," came Dr. Sargsyan's instructions over the radio. "Good, now sharpen the image. Stop. Save that image."

"Roger that," I responded, pushing the button to take a screenshot.

It was the first ever ocular exam in space using ultrasound.

Salizhan and I imaged each other's internal organs, bones, even teeth. This was a great demonstration that with a little training, astronauts could produce diagnostic-quality images. It was important research for long-duration spaceflight, especially when we go to destinations like Mars. It was also cool when Dulchavsky used the technique on the ground in developing countries. It was a great example of something developed for spaceflight benefiting people on Earth. He and I hit it off immediately during training and are good friends to this day.

During missions, NASA and Roscosmos monitor everything, including food consumption. Standard procedure is for us to scan the barcode on the container boxes (which contain mostly soft food packages) once they're empty. That way, ground control can determine how much food we're eating and how much we have left.

One morning, a month into our stay, the radio crackled to life. It was mission control calling.

"ISS, Houston. We've been monitoring your food consumption over the last few weeks. It has been unusually high. At this point, you have a significant shortage. We are evaluating and will be sending up a plan on rationing until the next resupply ship arrives in about six weeks."

What the hell?

Salizhan got the same message from the Russian control center. We met up in the Zvezda module.

"How could this be possible?" I asked. "We have *not* been eating more than planned!"

"Yeah, there's no way . . ." Salizhan said, as confused as I was.

It didn't take us long to agree about what we thought happened.

"They ate more food than they reported," he said.

"Why would they do that?" I asked in disbelief. "What harm would have come from doing an accurate tally?"

I went to the boxes.

"Salizhan, look at this." I opened one box, then another.

I realized that every single food container box—all the boxes that should be *completely* full—were only half full.

We'd never noticed before.

"You've got to be kidding!" he said, shaking his head in disbelief.

In zero gravity, the boxes have no weight. A full food container with soft food packages packed in as tightly as possible (to save space) has a certain weight. In zero gravity, if you remove half of those soft packages, the remaining ones expand and fill the empty space—making the boxes appear full. You would only know if the box was half full if you could feel how much lighter it was. Of course, we couldn't feel the weight of anything. I pushed the food pouches against the box, compressing them almost halfway across. *Oh my God.*

Padalka and Fincke had eaten WAY more than they accounted for. They even broke into Salizhan's personal "bonus" food containers. We did an inventory of our remaining food. We were nearly out of most of our favorites, and we'd only been in space four weeks.

Since the containers were only half full, it looked to our space agencies like we were eating twice the food we actually were. The meager amount of replacement food we brought up with us was woefully inadequate to make up for what was missing. It could have been an accounting error by the specialists on the ground, but Salizhan and I doubted it. If they had been given accurate details on what Fincke and Padalka ate, how could they make such a huge mistake? *Bastards!*

Sure, it likely wasn't on purpose. It's probably human nature to downplay and underreport when you take things or eat other people's food. But this wasn't like when you eat the rest of your sister's potato chips on Earth with a 7-Eleven down the street. We didn't bring enough food because Fincke and Padalka likely completely misrepresented what they snagged from our supply.

NASA scrambled to come up with ideas to get us through until a supply mission—scheduled for late December—arrived. They gave us a strict diet to eat about half of the calories we typically would. Fincke and Padalka left lots of sugary snacks and crackers, but we didn't have enough *real* food. We would be going on a supermodel diet, and we weren't happy about it. NASA came up with a few ideas, some of them just stupid. The worst being that we should stop exercising to burn fewer calories and take sedatives to slow our metabolisms. Can you imagine? In zero gravity, it's critical to exercise or your muscles and bones turn to mush. And how were we supposed to do quality work if we were on drugs?

A Russian media source found out our food was running out and called us "*obzhora,*" the pigs at the party stuffing our faces. NASA said they were looking into the mix-up but knew not to blame Salizhan and me. Bill Gerstenmaier, the ISS Program Manager, told a press conference, "This crew has not been eating excessively in any shape or form."

When I got a chance to talk to reporters at a press conference, I served the story straight: "The astronaut [meaning me] says US space agency officials believed there was more food on board than actually existed when the new crew took over in October, perhaps because the previous residents did not accurately inform the ground how much food they had consumed."

Padalka was a celebrated cosmonaut, so there was no way that the management was going to make him look bad or blame him for anything. This was the Russian way. Roscosmos blamed the people in the control center and said that they screwed up the accounting.

Back home on Earth, the local TV news interviewed my dad and asked how concerned he was about his son up in space without enough food.

"It's fine," my dad said. "He needs to lose weight anyway."

Thanks, Pop. We knew a supply mission was coming at the end of December, so we had about six weeks to ration food. If we really got hungry, we had a ton of crackers, desserts, and candy bars.

As far as meat, vegetables, and potatoes, we were limited. We had about 1200 calories a day of real food.

Salizhan and I tried to keep laughing. After sharing a measly dinner of one ration of chicken and noodles:

"Damn, that was good, Leroy, I'm so full I can't possibly eat another bite."

"Yeah, that was WAY too much food. I'm stuffed!" I'd say, and we'd both crack up.

To keep up our spirits, our friends at NASA and Roscosmos sent us jokes via email. We loved them and told everyone to keep them coming. A couple of the NASA nurses sent up one of the best ones: *What do you call two astronauts in dinner jackets? Hopeful.*

Soon, NASA's deputy chief astronaut Peggy Whitson, told everyone to stop. "There'll be no jokes," came the order to the teams. "We can't let any of this get out to the press."

Salizhan and I were mad. *You jerks. We're dealing with this with humor, telling you what we want, and you won't do it.* It was typical of NASA management. There was very little chance of the media getting hold of the jokes. They were going to our email addresses, not being read up on the radios. Even if they did get hold of them, so what? The Russians kept sending jokes. It fit with their culture to almost celebrate, or at least make heroic, our suffering. One of their favorite sayings was: "Life . . . It's a struggle."

On the morning of Thanksgiving 2004, I woke up imagining being back in my family home in Northern California during the holiday. My two sisters and I would sit at the kitchen table and peel off layers from the stack of dumpling wrappers. We'd grab a blob of pork and cabbage mixture with a spoon and drop it into the middle of the circle. Then we'd wet the edges of the wrapper, fold it in half, and pinch the edges to form an imperfect dumpling. Every holiday in Chinese homes is an occasion for *jiǎozi*.

Mom put them in water to boil, then got busy with the main course. She massaged a chicken all over with a specially seasoned salt rub that she made herself and baked to give it a roasted flavor. Inside the chicken, she stuffed sticky rice, *nuòmǐ,* made with soy sauce, seasonings, mushrooms, and sausage. The juices from the chicken morphed all the flavors into something even sweet potato casserole with marshmallows couldn't beat.

The memory made my mouth water and my stomach growl.

Salizhan didn't know it was an American holiday. We spent the day as usual. He worked on the Russian side of the space station. I was in the US lab. We were hungry and both a bit grumpy.

Before dinner, I went to my "rack," a phone booth–sized area that was my living quarters. I kept my personal stuff in a mesh sack clipped to the wall. My laptop stuck to a tray with Velcro, with an articulating arm like the ones photographers use attached to it. I'd been training with Salizhan for four years, so I knew his favorite space meals. A couple of weeks before, from our dwindling rations, I squirreled away some food for today.

I grabbed the Ziploc bag and translated over to the galley.

"I have a surprise," I say, pulling out the space delicacies one by one.

Sharipov's eyes got big, and he cracked a wide smile.

"Happy Thanksgiving!"

Thanksgiving dinner was a needed break from our low-cal diet. We only had a normal-sized meal, but it felt decadent compared to the last few weeks. Unlike on Earth, we didn't mix different foods together. In space, you eat sequentially because you don't want to have a lot of open stuff flying around. We sucked sauce out of beef steak packages then savored the irradiated meat. We slurped the Japanese curry and rice. Lastly, we added water to the Russian mashed potatoes with butter and onions—*delicious.*

The resupply ship came on Christmas night carrying a generous supply of food containers. We had to wait until the next day to do all the checks, make sure the seals were holding, equalize the pressure before finally opening the hatches. That big pile of food may not have been delivered by Santa and his reindeer, but it was the best Christmas present we could ask for.

PART FOUR
DESSERT

After Orbit Treats

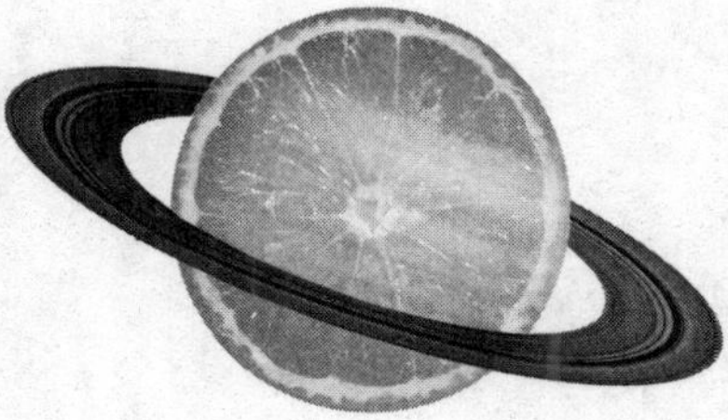

Q18: HOW DID YOU ADJUST TO "LIFE AFTER ASTRONAUT"?

AFTER MY LAST MISSION as International Space Station Commander, I left NASA in 2005. As you can imagine, in my mind, there was nowhere up from "astronaut" as far as careers go. I wasn't sure I wanted a full-time office job. Instead, I had my fingers in a few different pies. I became a consultant and got involved in a private commercial space company, plus a few technology startups. I did a little teaching and academic work, first at the Louisiana State University, then at the Baylor College of Medicine and at Rice University. But one of the coolest part-time gigs I got was in 2012 when I was invited to be on the safety advisory panel for SpaceX.

At the time, Elon Musk was known to be a quirky visionary. Love him or hate him (or more specifically, his politics), nobody can argue with his vision and successes. From

SpaceX to Tesla to companies like Neuralink, what many so-called "experts" claimed were impractical or impossible, he continues to steamroll forward. With no engineering degree, he can actualize anything without the constraints that most engineers have built into their mindset. Starting in 2010, SpaceX built Falcon 9 rockets and launched satellites for other companies and governments. But Musk knew that Falcon 9 wasn't big enough to get to Mars. So, he created what became the Starship program, the largest reusable rocket with a resusable winged vehicle. But he needed more capital, so he came up with a plan.

To fund his Martian dream, in 2019, Musk created a satellite network to provide fast internet coverage for the whole planet. Traditional satellite internet came from satellites in geosynchronous orbit, around 25,000 miles up. There's a significant time delay with the signal (hence the terrible service on your airline flight before the SpaceX system started getting adopted by the major airlines). But with low Earth orbit satellites, the constellations are around five hundred miles up instead of 25,000 miles, so the signal travels much more quickly. Musk called the satellite web Starlink, and it was a big hit.

SpaceX was not the first company to try to build a low-latency satellite internet constellation. However, Musk was able to leapfrog the other efforts, some of which had been in business for several years and only started launching a few satellites in 2025. He kept the majority interest in SpaceX so no board of directors or shareholders could crash out his dreams of going to Mars.

Once SpaceX landed a significant NASA development grant in 2012 to work on ferrying astronauts to and from the ISS, I was asked to be on their safety advisory panel. The panel included former NASA astronauts, program people, and engineers—including former astronaut Mark Kelly, who would go on to become a US Senator from Arizona and get into very public feuds with Musk over the war in Ukraine and other political issues. We were tasked with looking over SpaceX's shoulders as they got ready to fly astronauts on Crew Dragon and their Falcon 9 rocket. We met about three times a year, mostly in Hawthorne, California, for briefings from SpaceX senior engineers. We asked questions and toured the factory floor to see the flow of operations. We weighed in on the crew controls.

I flew their simulator several times and docked the virtual Dragon to the virtual ISS. The spacecraft was designed to do everything on autopilot—much like the Russian Soyuz and the Chinese spacecraft, *Shenzhou* (Dream Vessel). The Falcon 9 rocket first stage was being designed to be reusable, to come back and land at the launch site or on an ocean platform for reuse, which seemed bonkers at the time. (It happened successfully for the first time in 2015.) *Why hadn't anyone else designed a rocket like that?* It saves millions and millions in mission costs for each successive flight. Very impressive. But at the time, almost none of us in the business, including me, believed that they would succeed, at least economically.

Designing and building things to go to space is complicated no matter how great the engineering. There are always failures. On June 28, 2015, a Falcon 9 rocket, carrying a Dragon cargo capsule to the International Space Station, experienced a failure shortly after launch. At an altitude of about 150,000 feet, the vehicle disintegrated in a fiery explosion. Fortunately, nobody was hurt, and no damage was done to anything else. We were briefed by SpaceX on what their analyses showed. A helium tank mounted inside the liquid oxygen tank of the second stage broke loose during flight because a structural strut that held it in was made from substandard steel. Scary stuff. What if it had been a Crew Dragon with astronauts onboard? The abort system very likely would not have been able to react in time. Fortunately, Crew Dragon would not be ready to fly for the first time with astronauts for another five years, after much more flight heritage had been gained on the Falcon 9.

May 2020 was the first SpaceX flight test with a crew of two NASA astronauts on board. The company was running on a can-do attitude. SpaceX was acting like the first NASA Apollo engineers, working long hours and getting it done.

Q: DID YOU WORK FOR SPACEX?

I MET ELON A couple of times when I served on that White House space committee, then a few years later when he was getting an award at an event. I haven't spoken to him at length, but he's an interesting guy—and a little different—not in a bad way, but I could just feel his energy.

The astronaut who formed our safety advisory panel, Garrett Reisman, was a SpaceX employee and a friend. When he was going to move on from his job, he called me, letting me know that he was leaving his post at SpaceX as the Director of Space Operations.

"You should really take this job."

"Eh, thanks, but no," I said.

As I understood it at the time, SpaceX had standard employment packages for engineers and executives. There was no negotiation. *Here's your deal. You're going to get this much a year, plus some stock options.* It wasn't a lot for senior engineers, and there was no wiggle room. It was *take it or leave it. This is what you get.*

"Frankly," I told Garrett, "I need to make more money than that, and I don't want to move to LA. My kids are in school here. My family's here."

I was married at the time, and my twins, a boy and a girl who had become and are still my very favorite people in the whole world, were at the important developmental age of eleven. They were born in Houston and were growing up with their friends here. Uprooting us all at that point would have been tough on them.

"Besides," I said, "I don't think I can work for Elon. I think I'd get fired fairly quickly." It was just my gut feeling. I had been working for myself for about ten years at that point and I didn't want to go back to working for a large organization, or anyone, again.

Q19: WHAT'S HAPPENING WITH NASA TODAY?

ON JUNE 6, 2024, I was doing an on-camera interview with CNN from my home office. I'd done hundreds of these over the years and was comfortable being on TV and giving expert commentary. Along with hosts Boris Sanchez and Brianna Keilar, I watched NASA's live feed of the Starliner capsule approach the International Space Station. The docking was newsworthy because it had been so long in the making. It was the first time Starliner—designed and created by Boeing and years overdue—flew astronauts to space. The two-member crew, veteran NASA astronauts Suni Williams and Butch Wilmore, planned to stay for an eight-day mission on the ISS then return home.

The spacecraft floated through the dark, and with the help of thrusters to steer, lined up to dock with the space station. Starliner had experienced some leaky valves that caused several of the thrusters to fail, so this was a bit of a nail-biter. I'd had a docking emergency before of my own, with the Russian Soyuz spacecraft, back in 2004 during my ISS mission. So I had an idea of how Butch and Suni may be feeling.

CNN host Boris Sanchez came on.

"Leroy, as we're watching this unfold, take us inside the minds of these astronauts about to accomplish something historic."

I honestly didn't feel like this docking was *that* special, but I know how the media tries to hype up everything. At the time, I didn't know the full extent of the problems that they were having.

"Oh, absolutely," I said. "Butch and Suni are consummate professionals. They're very laser focused on the task at hand. Docking is one of those critical maneuvers."

A minute later, Starliner lined up and locked in.

"So, Leroy," Sanchez said, "put this docking into context for us, what it means for the bigger picture of space travel."

"What it means is that we now have a second commercial path to launching NASA astronauts to and from the ISS," I said. While I went along with their heightened interest over this maneuver, I had to put things into perspective. "Of course," I added. "SpaceX has been doing it for nearly five years."

A spacecraft docking to the ISS wasn't a new technological feat, *obviously*, and Boeing had been outrun by a much younger and faster company. But there was an upside. "NASA likes redundancy," I added. "Astronauts like redundancy. All engineers do. So, this would mean we have a second US path without having to go back to relying on the Soyuz vehicle."

We would learn nine months later from Butch and Suni, after their return, that it had been a lot more of a problem than the public had been led to believe. There was a (thankfully temporary) loss of full control of the spacecraft that made them wonder not only if they'd be able to dock to the ISS, but if they'd even be able to get back home if they couldn't.

As I signed off from CNN, I had no idea that Butch Wilmore and Suni Williams (two relatively unknown civil servants with cool jobs) would captivate America and become the most famous NASA astronauts in decades.

After they were safely on the ISS with the crew already there, the parked Starliner continued leaking from various valves. Boeing and NASA delayed Butch and Suni's return while they spent weeks troubleshooting and trying to understand and fix the problems. Finally, NASA decided the five-billion-dollar-plus (including development costs) Starliner was too risky to carry Butch and Suni home.

For the next nine months, other space professionals and I were endlessly pursued by media across the world to talk about the "scary fate" of the "stranded" astronauts. I'm not sure why the media was so fixated on this. On the one hand, astronauts like to be in space, so maybe getting extra time aboard the ISS could be a positive. On the other hand, the disruptions to their personal and perhaps also professional lives were a big deal—much more than missing a wedding or birthday back on Earth. But Suni and Butch were hardly in danger or suffering unduly. Headlines sensationalized the "abandoned" astronauts. *"Torment for families of two astronauts trapped in space . . ."* When Butch, Suni, and the two other ISS crewmembers posed for a Christmas photo, social media conspiracy theorists went into a frenzy denying Butch and Suni were even in space at all. *How could they have known to bring Santa Caps if they were coming right down??* (OMG, the Santa caps had been up there since before *my* ISS flight.)

Months in, Butch and Suni found themselves in a political tug of war when Elon Musk stated that he offered to send a dedicated rescue mission to "save" them with his SpaceX company, but that President Joe Biden's administration turned him down for political reasons. I had no reason to doubt that the offer was made, but it may not have gotten to the President, or he may have been advised against it because of the additional costs. The STRANDED ASTRONAUT story was the biggest space story in years. To me, it was ridiculous. Missions get changed up all the time for a variety of reasons, and there was no imminent danger.

Butch and Suni did get stuck longer than expected. That was true. In my opinion, their extended stay at the ISS was the result of a system that's become antiquated—even sloppy over the decades. In the old days, Boeing was an engineering-led company and was synonymous with quality and excellence. We engineers used to joke: *You can always tell a Boeing engineer because a Boeing engineer is wearing a belt AND suspenders.* They were double and triple dedicated to safety. Being conservative was how Boeing built quality into their airplanes and spacecraft.

More than three decades after they built the first stage of the Saturn V rocket that shot astronauts to the Moon, Boeing merged with McDonnell Douglas and reportedly morphed into more of a bottom-line, MBA-led company. Profitability and profits became the most important thing. The culture shifted at the top. People who worked for Boeing for decades blew the whistle on unsafe practices to politicians and the news media that resulted in three fatal airplane crashes with their 737 MAX planes in 2018 and 2019. Investigators found that Boeing had done questionable design work on a MAX flight control system and covered it up (one test pilot email stated that he used a "Jedi mind trick" to get it past the Federal Aviation Administration representative, it was reported) to save time and make more money. The company got away with it, several news outlets reported, because the FAA was so familiar with the contractor that federal oversight workers had become complacent and didn't pay enough attention to what was happening with the MAX airplanes.

Q: WHY DID NASA NEED A NEW VEHICLE ANYWAY?

IN 2011, NASA LANDED the Space Shuttle for the last time and retired the Shuttle program. The Shuttle's final purpose was to build and service the ISS, and construction was finished. Also, the end of the program was sealed after the 2003 *Columbia* accident. In the beginning, the estimated risk of loss of vehicle (LOV) and/or loss of crew (LOC) was one in ten thousand. After the *Challenger* and *Columbia* accidents, the real-world risk turned out to be one in sixty-seven. Not good. In 2004, President Bush decided that the program would end after the last ISS construction mission in 2011. At the same time, he announced the beginning of a new exploration program called "Constellation" to return humans to the Moon and land astronauts on Mars. *Sound familiar?*

By the time Barack Obama was elected President in 2008, the Constellation program was in big trouble, in funding and estimated timing and hardware development. Orion was the name of the new capsule being developed to fly astronauts to and from space. Because of all the issues, the new administration created the "Review of US Human Spaceflight Plans Committee" out of the White House Office of Science and Technology Policy with aerospace legend Norm Augustine as its chairman. The two former NASA astronauts on the committee were Sally Ride, the first woman to fly into space for NASA, and me.

As we began our work, it was quickly clear that there was no way the Orion spacecraft or the rocket to lift it (both under development) would be ready to fly anywhere near 2011. We were completely dependent on the Russian Soyuz to bring all astronauts to the ISS after the *Columbia* accident until the Shuttle returned to flight in 2005. If nothing was done, we would be dependent on them again in 2011 when the Shuttle was to be decommissioned.

In 2006, NASA started a program to help commercial companies develop the capability of flying cargo to and from the ISS because Orion was not designed to carry much cargo. Since nobody could predict when Orion would fly astronauts—if ever—our committee suggested expanding the cargo program to help these companies develop a spacecraft that could also fly astronauts to and from the station.

NASA put out a first request for proposals for the Commercial Crew Program in August 2009. The requirements were to be able to fly a crew of four to the ISS, dock, stay there for at least 210 days, then bring the crew back. After four years of NASA-funded development by a handful of companies, NASA awarded final contracts to two companies in August 2014 to build and fly their spacecraft. These were the high-dollar contracts, which were prepayments for future services, not development grants. This was a new way of doing business. It saved significant taxpayer dollars and allowed the companies to sell their services commercially to anyone, with NASA as the anchor client. One was SpaceX.

At that time, SpaceX's main revenue stream was building Falcon 9 rockets to launch satellites for government agencies and commercial companies. In May 2012, SpaceX also started delivering cargo to the ISS in a spacecraft called Dragon 1. For SpaceX, getting astronauts to the ISS would be a matter of modifying their Dragon vehicle for humans—no small feat, but they'd been working on that from the beginning.

NASA didn't have full confidence in SpaceX (they were the new kids on the block), so they also gave a contract to Boeing—for almost twice as much money because Boeing is so much more expensive (they have the highest overhead rates in the industry). NASA's thought process was: *We like SpaceX, but we know Boeing can do it. They've been in aerospace for so many decades. They've created all these different vehicles, rockets, everything.*

While Boeing stumbled along and their Starliner vehicle had setback after setback, the first SpaceX Crew Dragon (which they named *Endeavour* after the last Space Shuttle) took NASA astronauts to the ISS on May 30, 2020, on its first crewed operational flight test. It was a big success. Crew Dragon then remained the only American vehicle flying humans to space until 2024 when Starliner finally got cleared to launch Butch and Suni on its first crewed flight test.

When I heard about Starliner's failures, I just shook my head. Boeing had given NASA a 2017 launch goal. Instead of delivering on time, they missed their *own* deadline. Boeing management and engineering were in disarray. The parachutes used to land the Starliner had issues. The cooling system needed too much water. Valves corroded. Software glitched and sent the Starliner into the wrong orbit during an uncrewed flight test. Even the Atlas V rocket was found to have a faulty valve in its Centaur upper stage, delaying

Butch and Suni further. *Boeing, you can't even get your rocket right?* Finally, when Starliner was about to be delivered to NASA, someone discovered that the tape they used on the wire bundles was flammable. *Seriously?* They had to go back, take all the wire bundles out, undo the tape and re-tape everything with a suitable tape.

Boeing tried to brush off the bad valves with an excuse about being in a humid environment down at the Cape. But we'd always launched from the Cape, so they're used to operating in that humid environment.

Finally, NASA cleared the launch—even though Starliner had two leaky valves. There were enough redundant valves and thrusters that NASA and Boeing decided it shouldn't be a problem. But in flight, when Butch and Suni were on their way to the space station, more valves underperformed, causing anomalies with the reaction control thrusters used to maneuver the spacecraft in orbit.

The situation was much more critical than NASA reported to the public at the time. Butch was flying manually and lost enough thrusters that the Starliner controls were sluggish, and he couldn't move the spacecraft forward. There was real uncertainty of whether Butch and Suni would be able to dock to the ISS, and worse, make it back to Earth if they couldn't reach the station. Heroes in the Mission Control Center came up with the idea of resetting the failed thrusters by essentially rebooting the computer. After a breath-holding moment, two of the four failed thrusters came back online. A second reboot recovered another one, and Starliner was finally able to dock.

After a bunch of tests in space, but without fixing the docked spacecraft, Boeing felt that Starliner was safe to get Suni and Butch back home. We astronauts were not as con-

fident. Ultimately it was NASA's call. "The decision to keep Butch and Suni aboard the International Space Station and bring Boeing's Starliner home uncrewed is the result of our commitment to safety: our core value and our North Star," NASA Administrator Bill Nelson told the media in a news conference on August 24. I thought that was absolutely the right decision. Why risk their lives on a vehicle when you don't fully understand what the problem is? If it had been me on that flight, I would not have wanted to get back into Starliner after coming that close to losing control of the spacecraft and almost being literally stranded in space.

There was an easy and obvious solution anyway. Three NASA astronauts and a Russian cosmonaut were set to travel on SpaceX's Crew 9 in September 2024. The four would replace Crew 8 that had been up there for six months, then be replaced by Crew 10 in February. NASA decided to kick two astronauts off Crew 9 and send up Dragon with two empty seats for Butch and Suni.

I wasn't worried about Butch and Suni having an extended mission. They had both flown long-duration missions before. They know the drill. They know how to operate. They know where things are and what's what. They were able to easily slip into that role. But my heart sank a little for the two astronauts who were cut from Crew 9. Imagine getting ready to go fly your mission. Then weeks from launch, after training for years, NASA says, "Sorry, guys. We're kicking you off the flight." *Ouch.*

I even brought it up in some interviews. *What about the people who got left behind?* To me that was the big human-interest story. One of the two would be flying her first mission. But the media couldn't care less about those two stuck here on planet Earth.

I did feel for Suni, though. She was assigned to the Starliner years earlier. It would be her last mission before she retired, so she was happy to get to fly on Boeing's new vehicle. We were all sure it would go before Dragon. Suni, a Massachusetts native, bought a place in New England to retire with her husband after leaving NASA, then she delayed her departure from NASA while watching SpaceX fly the Dragon over and over. Finally, she took off on the Starliner only to get stuck on a full long-duration mission—plus some (most long-duration missions are around six months, not nine).

Starliner came back by itself, entered the Earth's atmosphere and landed in the New Mexico desert where it was supposed to. Butch and Suni were added to the next space station crew that was launching in September on SpaceX's Crew Dragon. That crew planned to return in February. How Butch and Suni were coping was all any news people wanted to talk about, but in my mind, the media was missing the bigger picture. The entirety of space travel for humans was changing in new and exciting ways we'd never seen.

Q: YOU SAID THE MEDIA WAS MISSING THE BIGGER STORY. WHAT DO YOU MEAN?

THERE ARE TECHNOLOGICAL, philosophical, and bureaucratic things changing space exploration on a scale not seen since the 1960s. First, while every mainstream reporter was consumed with Butch and Suni, SpaceX pulled off the biggest coup in decades, and almost none of the mainstream media even noticed. As part of Musk's continuing push to get to Mars, on October 13, 2024, SpaceX launched the fifth Starship flight test. (The booster of Starship, called Falcon Super Heavy [FSH] is the most powerful rocket ever built and flown.)

In what looked like an AI-generated fantasy, the FSH, after spending most of its fuel getting the Starship high and fast enough to get to space on its own rocket engines, dropped back toward the launch pad with some of its engines relit and throttling. Then it maneuvered into position and hovered over the pad to be embraced by a capture mechanism with two arms nicknamed the "chopsticks." It was incredible. SpaceX proved that they could softly capture the FSH after launch, which meant they could refurbish and reuse it, just like they had been doing for many years on a routine basis with the much lighter Falcon 9 that self-lands with pop-out landing legs.

It also happened just weeks before the 2024 presidential election, and Elon Musk was deeply entrenched in American politics. Musk aligned himself with Donald Trump in the most polarizing campaign in recent history. With Musk becoming one of the right hands of the American executive branch after Trump's win, the case of the two "stranded" astronauts became very political.

While SpaceX gained more momentum in 2025 in both private and government spaces, NASA moved like a giant sloth. In 2009, the Constellation program was in a lot of trouble in terms of technical, budget, and schedule issues. Cost overruns were horrendous. The Obama administration canceled Constellation in 2010, then brought back the Orion spacecraft and created a rescoped, smaller rocket program called Space Launch System. Moon landings were scrapped (President Obama's chief of staff, Rahm Emanuel, pointedly told Norm Augustine in 2008 that the Moon was off the table because it was Bush's program). The goal became "visit an asteroid and then move on to Mars." One main problem was that there was no increase in funding on the table.

During President Trump's first term, he renamed the exploration program Artemis and turned the focus back toward the Moon. By 2025, the Orion spacecraft, built by Lockheed Martin, still hadn't flown astronauts, but might in 2026.

It's mind-boggling when I look at the history of NASA. The agency got us landing on the Moon just under eleven years after its inception—with 1960s technology! We created rockets, launch pads, mission control capabilities, and a robust team of engineers and well-trained astronauts. NASA's current human space exploration program, in one iteration or another, has been going on for twenty-one years, and we haven't launched a single astronaut into space on that program.

It puts into perspective how bloated, bureaucratic, and inefficient NASA and the traditional contractors have become. Everybody was on board during Apollo in the early days of NASA. Boeing and North American Aviation (which became Rockwell, which became part of Boeing), Lockheed, Northrop Grumman—were pulling in the same direction. Both sides of Congress were gung ho for what we called "manned" space missions back then. *We're in an existential race against the Soviets!* Then, when the Apollo program was canceled after Apollo 17 (there were to be twenty Apollo missions), the contractors realized that once they succeeded, the money ended. Lesson learned: *Don't let the mission end. Let's keep it going. Rescope. Stretch it out.*

Q: WHAT IS THE BIGGEST CHANGE IN SPACE TRAVEL AND IN SCIENCE THAT YOU ARE SEEING NOW?

SINCE THE BEGINNING OF HISTORY, wealthy people in societies have funded science and innovation. Inventors of all kinds turned to private funders and philanthropes to fund new technology in medicine, energy, transportation, art, and architecture. With NASA losing momentum and government funding, things today appear to be moving in that direction.

Enter young space junkie Jared Isaacman. In 2021, tech billionaire Isaacman bought his first mission from SpaceX. Isaacman used the mission to raise money for children's cancer research at St. Jude's hospital with his own $100 million donation and selected individuals for the three other seats, based on his evaluation. It was the first time that a spaceflight didn't have a single government or former government astronaut on board. It was completely "Space Agency" free. SpaceX trained the private citizens to operate Dragon, launched the crew on their Falcon rocket, and conducted mission control from its California headquarters. Isaacman and crew flew in orbit for three days. On his second flight in 2024, Isaacman and SpaceX tested the company's new prototype EVA suit—so streamlined (but also much less capable in the prototype configuration) compared to our NASA suits. Isaacman and Sarah Gillis, a SpaceX engineer, peeked out of the hatch like Punxsutawney Phil. All went well, and it appears spring is coming soon for SpaceX and other companies in the private and commercial space travel arena.

In March 2025, Butch Wilmore and Suni Williams returned to a different world than they'd left nine months earlier. The future of space would no longer be determined only in the halls of NASA and the giant factories of legacy contractors. The days of sending humans to space only on government programs like NASA's were in the past. Money—and the egos of space-loving billionaires—have become the new driving force taking humans into orbit and beyond.

Q20: WHAT IS THE FUTURE OF HUMAN SPACEFLIGHT?

GEORGE JETSON WAS AHEAD of his time.

In the 1962 cartoon series, *The Jetsons*, when his flying "space car" ran out of fuel, George filled up with "two bucks' worth of high-octane pellets" at a self-service depot in space—no need to go back to Earth to gas up. More than six decades later, SpaceX and other companies like Blue Origin are working to implement that idea—one that promises to take us way beyond where we've flown before.

The thing that's always limited how far we can go is fuel. The rocket has to burn so much propellant to get us to space that once we're up there, the tanks are usually on empty. But like in Jetson's Orbit City, if you pre-position fuel depots in orbit, your spacecraft can rendezvous, dock with a depot, refuel—and you're off—to the Moon, an asteroid, and beyond.

According to Elon Musk, some of SpaceX's Starships will serve as orbiting fuel depots, with demonstrations starting in 2026. Other Starships will fly up, refuel, and go on to their destinations. When an outpost tank is almost empty, it will deorbit (come back to Earth), get refurbished, then launch back to space with a full tank of rocket propellant (minus what is needed to get into orbit). The concept is brilliant (the idea has been around since the 1960s), but refueling in space is challenging. How do you transfer propellant in zero gravity? Can you keep oxygen cold enough (-279 degrees Fahrenheit) to stay liquid? How hard is it to dock one Starship to another? It all seems so sci-fi, but I'm sure that smart engineers and aerospace designers can make it work.

As mentioned earlier, SpaceX isn't the only company building fuel depots in space. The US Space Force plans to test an orbital fuel post in 2026 that's being built by an aerospace startup called Astroscale. Jeff Bezos's company, Blue Origin, is developing Blue Ring, a vehicle that they say will not only fuel other spacecraft but be a whole orbiting service station.

The SpaceX Starship system is comprised of the rocket booster (Falcon Super Heavy, FSH) and the winged vehicle launched by the FSH (Ship). While SpaceX engineers have successfully grabbed the FSH several times in the "chopsticks" and demonstrated reusability by launching a recovered and refurbished FSH, Ship itself hasn't yet had a fully successful flight test. Once the issues are worked out, and Starship gets certified for carrying humans and launching things to space on the regular, it will change the entire game. It's like a giant, fully reusable Space Shuttle on steroids—able to carry 150 metric tons of payload to orbit, and one day, according to Musk, a version will fly one hundred humans at a time on a journey to Mars after making a pit stop in orbit.

Q: WHAT IS THE LIKELIHOOD THAT WE CAN ACTUALLY LIVE ON MARS?"

ELON MUSK *ALWAYS* SEEMS too optimistic—but he hopes that by 2029, Starship will reach Mars, where he plans to go and build colonies. He says it's something he's wanted to do since he was young. But it's not an easy task, which is why NASA has been talking about it for decades yet has never made it happen. Depending on how the planets are lined up (Earth's distance from Mars changes over time because of the difference in our orbits), round-trip travel on Starship with a few months' vacation on the red planet could take two years or so.

Maybe there are some young Leroy Chiaos out there in second grade thinking, "Wow, I want to go to Mars." I get it.

But visiting is one thing, living there is another.

Say a group of people today, a few families, have had enough of the hustle and hassle of planet Earth in the twenty-first century. Musk is betting they'll want to take a chance on Mars. To me, even as a huge space fan, Mars is a hard sell. For one, you can never be outside without a fully functioning spacesuit because the atmosphere is mostly carbon dioxide and would suffocate you immediately. The pressure at the surface is only six one thousandths of Earth's pressure at sea level. Mars's extreme temperature changes cause massive dust storms on the regular. If you're living in a giant terrarium (you'd have to be living in something like that), the dust storms can mess with all the electronic systems keeping you alive. And if you're looking for a room with a view, you won't even see the Martian surface or the sky when dust storms kick up. Another downside? Mars, unlike Earth, doesn't have a global magnetic field (magnetosphere) or thick atmosphere to protect life from radiation.

Musk most likely will choose the northern hemisphere for his outpost, where summer is longer (with temperatures in the high 60s Fahrenheit), and winters are shorter (-195 degrees Fahrenheit). Musk isn't known for having luxurious taste, so I'm not sure how the digs would be—likely more McMurdo Station in Antarctica than The Ritz Carlton in Paris. And remember that delay when you first got a cell phone in the '90s and called long-distance? Communication back to Earth would take about fifteen minutes on average, depending on the relative positions of the two planets. *Can you hear me now?*

I'm not sure how SpaceX makes money betting on Mars. How many people would take a tourist mission to the red planet? How are they filling the seats and paying the bills to run the outpost and the Starship? Even if SpaceX has customers cashing out their life savings and dropping a million dollars or so for the opportunity, it's hard for me to see it being a revenue positive venture.

Q: DOES SPACEX HAVE ANY COMPETITION?

EVEN THOUGH HE GOT a small head start with Blue Origin, Jeff Bezos quickly fell behind Elon Musk in the New Space Order. The billionaires feud in public, but I think their missions are complementary. Musk is focused on Mars, and Bezos is focused on the Moon and what we call cislunar space, the area between the Earth and the Moon. Bezos wants to build infrastructure in orbit. He wants to build factories in orbit and on the Moon with the idea to move manufacturing off the Earth.

Bezos is obviously a smart guy too, but I'm thinking: *How do you get factories up there and make products?* You've got to build your factories, ship raw materials up there, make your products, and bring them back. How is that going to be affordable?

Q: WHAT'S HAPPENING WITH THE RUSSIAN SPACE PROGRAM?

WITHOUT THE UNITED STATES, I'm not sure the Russian space program would have survived after the Soviet Union disintegrated. Even with our subsidies, Roscosmos hasn't come very far from their 1960s technology. I was shocked by their antiquated factories when I visited in the '90s, and not a lot of money has gone into improving those infrastructures. The electronics and communication systems have been upgraded, but they're using the same basic spacecraft—the workhorse Soyuz—the same rocket engines, same everything.

Aboard the ISS, there are some real issues. The Russian modules are only slightly evolved from the ones used on the Russian *Salyut* and *Mir* stations. Those original modules were only designed for a six-year life. *Mir* was having serious problems after twelve years of operation and was finally brought down after fifteen years. The ISS versions have currently operated for over twenty-five years, with the intent on lasting to thirty years. The age is showing. An air leak was discovered in the Zvezda service module in 2019, and in 2021, cosmonauts found tiny cracks in the metal that they've been patching with epoxy. With the other Russian modules built from the same metals and same basic designs, it's a valid question whether the ISS is going to make it another four years before it's set to be decommissioned in 2030. Is there a possibility of some kind of catastrophic failure? It's a little dicey, but I don't think so.

Vladimir Putin came to power in 2000 and has been in office ever since (except for 4 years when there was a puppet government and Putin pulled the strings behind the scenes). He's never been a major space fan or huge champion of Roscosmos, and although he claims to support building of a new Russian-only space station, the amount of rubles it would cost has kept the project on the drawing board.

With the Russia-Ukraine war isolating Russia from the West, the political umbilical cord that held NASA and Roscosmos together for the last quarter century is slowly coming undone. While I'm the first to admit that I never wanted to work with "the enemy" back in my early years, it makes me sad to say goodbye to what's been a truly wonderful partnership. Most of us at NASA have become lifetime friends with our Russian counterparts. We always look forward to meeting at conferences and other events. Politics never got in the way. We had spirited debates about world events, but we were always respectful and friendly.

Q: HOW ARE CHINESE ASTRONAUTS FARING IN THIS NEW SPACE RACE?

THE CHINESE HAVE BEEN making great progress, advancing steadily over the years in human and unmanned spaceflight, and—unfortunately for us and the other nations with space programs—all on their own.

After I left NASA in 2005, I made several trips to the Astronaut Center of China in Beijing. I met several of their astronauts, and I especially liked connecting with the first Chinese national astronaut, Yang Liwei, and their first spacewalker, Zhai Zhigang. Most flyers are cut from the same cloth all around the world, so we had instant connections, just like I experienced in Russia. I also got to know the deputy chief designer of their space program, Chen Shuangang. It was cool to use my Mandarin in a professional setting. *Thanks, Mom and Dad.*

Hu Jintao, the president of the People's Republic of China at the time, was a big proponent of international collaboration in space, and the Chinese wanted to be part of the ISS. The Europeans, Canadians, and Japanese were up for it. Then in 2011, West Virginia Congressman Frank Wolf, backed by several other Republican representatives (including John Culberson, my own representative in Houston), led the charge to exclude China.

Even after Congress blackballed the Chinese from the ISS, China still wanted to collaborate, and so did President Barack Obama and NASA administrator Charlie Bolden. Charlie asked me to help make connections with China's manned space office. I was fired up to help, hoping to see the same positive developments as our cooperative efforts with Russia had yielded. But there was so much pushback by Washington to do anything China that the State Department never issued visas to the Chinese astronauts and almost never to most of the aerospace engineers and space medicine specialists.

By the time Xi Jinping took over as president of the People's Republic of China in 2012, China no longer wanted to collaborate with NASA. In 2021, China quietly launched their own space station they called *Tiangong* (Heavenly Palace). Their *Yuhang yuan* (the Chinese name for astronauts which translates to Space Travelers) fly six-month long missions on the station and conduct biotechnology, material science, fluid physics, and a lot of other research, just like we do aboard the ISS. They regularly conduct EVAs and operate robotics from inside their station. In retrospect, perhaps China's intention with their earlier overtures to the West was to learn as much as they could from NASA (the Russians had already done all they could to help China's program) to move more quickly.

Private Chinese company Deep Blue Aerospace is trying to do what SpaceX has done and build reusable rocket stages. In the first test of Nebula-1, the rocket came down, its four landing legs popped out just like on the Falcon 9. Nebula-1 looked ready for a perfect landing, then it dropped to the ground harder than it should have, collapsed the legs, and blew up on the pad. They'll get it right eventually. It took a lot of tries and explosions before SpaceX started nailing it regularly.

Xi Jinping's goal, which is not a secret, is to become the world's number one superpower—and that includes in space. In 2023, the China Manned Space Agency (CMSA), announced that they'll land astronauts on the Moon by 2030 in their own Chinese designed and built lunar lander. For China, this would be their first time to the Moon.

Q: AND WHAT'S GOING ON WITH NASA AND THE AMERICANS?

AT THE TIME OF this writing, NASA plans to fly the Artemis II crew in early 2026 and take four astronauts around the Moon. Okay . . . *but* . . . critics would be correct in saying, "Hey, we flew the first humans around the Moon in 1968, and now we're doing it again? How many years later and at what cost? How much have we invested in this program already?"

There are rumors that the Space Launch System (SLS) rocket might get canceled after we return astronauts to the Moon, or even before. By NASA's own estimates, they're spending $2 billion per rocket, not to mention development costs (it's over $4 billion each with those costs included). And after launch, the entire rocket gets thrown away. Then they commission another one from the legacy contractors. This is the old way. NASA didn't consider we might be able to reuse rockets until SpaceX came along. With Starship expected to be so inexpensive to operate (it is being designed to be fully reusable), how can we justify to US taxpayers that we're going to go with the old-school way? It's difficult to see how the Artemis program can survive in its current form.

However, NASA has awarded contracts to develop human-rated lunar landers to SpaceX and Blue Origin, and the acting NASA Administrator, Sean Duffy, has announced plans to land a nuclear reactor on the Moon, specifically before China and/or Russia does. We are focused there in the near term, with the longer-term goal being to land humans on Mars. This aligns with both companies. Blue Origin is focused on developing cislunar space, and SpaceX on developing Mars missions. I am hopeful that NASA will expand its commercial partnerships with both, to leverage the nimbleness and technical innovations from the commercial side. NASA will help with training and operations—because there's no organization that can run a space mission like NASA.

So maybe we can stay on top. Just like when I was eight years old, the thought of America giving up our number one status in space makes me feel awful. But I have to be a little bit circumspect at this point in my life. All great nations and civilizations eventually fall. There was a time when Portugal ruled the world. They established colonies almost everywhere. Then Portugal got overtaken, and countries like Spain, France, and England had their turns at or near the top. Before that, Rome was everywhere. All these great civilizations fell. We're a young country, but history tells us that it would be unrealistic to say we'll be leading the world for decades and centuries to come. Still, I hope we do.

Q: WITH ALL YOUR TIME IN SPACE, DID YOU EVER ENCOUNTER ANY EXTRATERRESTRIALS?

THROUGHOUT MY ASTRONAUT CAREER, I saw a lot of weird physical phenomena that played tricks on my eyes. Each time there was a valid explanation in the physical realm. It wasn't until August 14, 2024, when I was flying my 1979 single engine four-seater over the Texas panhandle, that I came across something I can't explain. Two shiny, smooth metallic spheres, stacked vertically with no visible physical connection between them, zoomed past my left wing just about twenty feet below me. There were no wings, no visible means of propulsion and no sound.

Part of me wanted to dismiss it as some visual aberration. It was surreal, but I was in clear weather on a bright, sunny afternoon with excellent visibility. So I knew it was real. I didn't call the air traffic controller (I was on an instrument flight plan), who would have alerted me if he had had the flying object on his radar. When I landed, I didn't report it to the Federal Aviation Administration. I didn't want people to think I was nuts. For months, I told almost no one. Eventually, I mentioned the weird sighting in a media interview, and the story got picked up. Then the FBI got in touch with me and wanted to interview me. Sightings of strange flying objects had become more common, and the feds were taking them seriously. Over a Zoom call, the agent asked if I thought it was a drone or some wildly advanced foreign technology. I told them I had no idea. It didn't look like anything I'd ever seen before, and it didn't move in a way that seemed physically possible. He asked if it could have been two metallic balloons. I considered that, but they seemed rigidly attached, and our closing rate was too high for balloons just floating around at nine thousand feet. These spheres were traveling in the opposite direction with speed similar to my airplane.

I got some serious validation in September 2025 when Congress held a public hearing focused on Unidentified Anomalous Phenomena, UAPs, which is the new official name for UFOs. Missouri Congressman Eric Burlison played a classified military video recorded by a Reaper drone over the Red Sea. In the footage, a missile is fired at a shiny round object zooming through the sky and makes a direct hit. The object doesn't explode and continues its course like nothing happened except a little bump. I watched that clip, trying to square the laws of physics with what I was seeing. The missile hit the target, but the object, whatever it was, appeared to be nudged off course momentarily, deflected the missile—and kept flying. Why didn't the missile at least explode?

The panel that day included Air Force veterans, Navy personnel, and investigative journalists. Serious people are paying attention to UAPs: defense officials, physicists, engineers, people who've spent their careers in technology and science. Some of them are starting to say out loud what used to be whispered in hangars and labs. We could be seeing something beyond our understanding.

In my opinion, what I saw that day in 2024 was most likely a supersecret man-made machine made in the USA or another world technology player. But if it was one of ours, why wasn't it in a military operating area (MOAs are airspace reserved for military operations)? I don't think my UAP sighting was alien. But I do absolutely believe that there's other life out there. Statistically, with the vastness of the universe, there must be what mathematicians call a "countably infinite number" of Earthlike planets that contain life as we know it. It's almost a certainty. The issue, I believe, is that we are too spread out to ever find each other. It's fun to think about though, and I still don't really have a good guess on what it was that almost hit my airplane on that clear sunny afternoon at nine thousand feet.

EPILOGUE

ALL GOOD THINGS

SPACEFLIGHT IS A MAGICAL EXPERIENCE. From 250 miles above, the Earth is a living kaleidoscope of blues and greens. Clouds, weather systems, and oceans spiral and flow in constant motion. The Moon on the other hand, is cold, gray, still, but beautiful in its own way. Looking into deep space reveals a multitude of stars shining brighter than you could ever see from a mountaintop on a clear night. When I flew into space on my first mission, it was the most profound thing I had ever experienced. I cherished the moments. I don't think anyone experiences spaceflight without some intense introspection and wondering, *Why are we here?*

At the time, I didn't realize there would be something even more profound and joyful for me—the birth of my twins, Henry and Caroline, in 2006. I'd watched my sister raise my two nieces, so I thought I knew what having kids would be like. I could not have been more wrong. When I saw my children for the first time, I felt an instant emotional bond and instinct to protect them that I knew would last forever. Although quite different, the emotions were similar in some ways to viewing the Earth from space for the first time: a mixture of awe, love, and protectiveness.

Henry and Caroline are adults now. They certainly presented me with some challenges over the years, but they also gave me many moments of intense happiness. They will always be my favorite two people in the world. My dad was hard on me growing up, and my mom went along with him. I don't begrudge them at all. Their strict discipline made me who I am. I must admit that it did cause me to overcompensate a bit with my own kids. I was not as hard on them, and I often gave in to things they wanted. I hope I didn't do them a disservice.

The human experience comes with ups and downs. You must have the bad or you don't appreciate the good. My dear friend Debbie Gary taught me that. She is a star in her own right. She was the first woman to fly on a professional aerobatic team. Over several decades, Deb flew in innumerable airshows in a variety of airplanes and has received many honors and accolades. She has the most amazing positive attitude and perspective on life. In a way, her approach is similar to the somewhat fatalistic attitude of Chinese culture. You work hard and strive to reach your goals and dreams, but what will be, will be. You take what comes with a positive attitude.

I was unhappily married for many years. I was off balance emotionally as I worked harder and harder to support my family to the expectation that was set on me, while becoming systematically more and more isolated from Henry and Caroline. My divorce was contentious and difficult. Some challenges and conflict continue because we share our children. But those difficult times made me so appreciative of finding Jami, the love of my life, who is a very close second to my kids on my favorites list.

On the intellectual side, studying engineering was the hardest thing I've ever done, especially earning my bachelor's degree. The gift of that effort was a deep understanding of how the physical world, and to a lesser degree, the universe, works. (The more we learn about the universe, the more we realize how much we don't understand.) Engineers see the world in black and white. We believe that everything should stand on merit. There are boundaries and silos that must be respected. That's why I was so incensed when President Clinton used NASA as a tool of his foreign policy to keep Russian aerospace specialists from working for adversaries like Iran and North Korea.

My good friend David Warden helped me see the light, or rather, see the *gray*. David started life as an engineer but spent the bulk of his career as a trial lawyer. He explained that in the law, there is never black and white, only shades of gray. Just because you're right doesn't mean you will win. Just because something is unjust doesn't mean you can keep it from happening. He opened my eyes to realizing that the whole world is gray. This helped me finally accept that politics are what ALL decisions are based on. Whether in the macro scale among nations, or the micro scale among family and friends. Almost everything is negotiable. Diplomacy is key.

For the handful of billionaires and influencers who have either bought their way to space or had it gifted to them, I'm not a hater. I am glad I was a professional space flyer, but it's not for everyone. This was my passion, my dream. I did it for myself, and I have no regrets. These new civilian flyers are amateurs, but they're helping shape the future of space exploration. I believe that the more people who get the experience, the better place this planet will be.

Why spend money on exploring space when there are a million things to fix right here on planet Earth? (*Note:* The number may surprise you; historically, it has mostly only been around one percent of the annual federal budget.) The practical answer is the technical spinoffs, the science we learn, and the tools we create. *But the real point?* Humans are motivated, curious beings. We need to know what's on the other side. We need to strive, to achieve. There is no greater unknown than what happens at the outer boundaries of the universe, and even at the outer boundaries of our own solar system. As humans, we need to have patience to make these discoveries—our universe doesn't bend to the scale of our lifetimes. About a decade ago, NASA's Voyager probes finally flew past the heliopause of our sun after being launched about a half century before.

Just as life ebbs and flows, so does human space exploration. NASA is not the agency it was in the 1960s during the Apollo program, but the commercial side is having enormous success. More commercial-NASA partnerships are coming online, and the future looks exciting. We will get humans to Mars. I am confident of that. Would I want to go? To explore and come back, for sure. But I wouldn't want to live in a Martian colony. Life here on Earth is so much better.

What I missed most during my long mission (besides family and friends) was nature. I missed seeing birds fly and hearing them chirp. I missed squirrels chasing each other up trees. I missed warm sunshine on my face. After the rescue crew opened the spacecraft hatch post-landing, I smelled muddy grass. A flood of Earth memories came into my mind, and I felt in my soul that I was home.

Later that day, I arrived back in Star City. My friend snuck into the quarantine area. With a big smile, he pulled out homemade chili dogs and bottles of beer from his coat pockets. The first bite reminded me how much better food is on Earth.

As much as I love exploring space in person, at the end of the day, life for me is all about the people I'm close to and the things we do together. In 2017, we lost Mom after a brief illness. I didn't know how to mourn her for almost a year. Then I spontaneously started making the dishes she used to make when I was young. I brought Henry and Caroline into the kitchen to pinch the *jiǎozǐ* wrappers and mix the pork filling. The smell brought back wonderful childhood memories for me, and we savored the flavor of our Chinese dumplings that tasted *almost* as good as Mom's. Henry and Caroline couldn't care less about having dinner with an astronaut, but they were over the Moon tasting Grandma's dumplings again.

★★★★★

ACKNOWLEDGMENTS

IN EVERYONE'S LIFE, there are moments and people that one will always remember, that influenced and guided the path forward. For me, it began with my parents, and my sisters, Judy and Sandy. They always believed in me, even in the moments when I doubted myself. Some people never knew how much they helped to shape me. My first-grade teacher, Miss Link, was my first love. I looked up to her and endeavored mightily to do well in class to earn her approval. She unknowingly taught me the value of hard work and focus. My after-school nanny, Nicey, was perhaps the kindest person I've ever known. She showed me the importance of being good to others. The first astronauts and the Apollo Moon landings paved the way for my generation. These astronauts were my boyhood heroes, and their exploits made me dream of one day following in their footsteps. Years later, I was fortunate enough to meet a few of them in person after I had been selected by NASA. Most of them likely never thought about how much influence they might have had on some young people like me. I am, of course, grateful to all these people, and others like them who have not been mentioned by name. And now, on to the team.

Thank you to my friend and manager, Glenn Schwartz, and Peter McGuigan, founder of Ultra Literary, for bringing the ideas and creating the team to make this book possible. Thank you also to Victoria Bruce, a talented writer and kind human being, who was able to capture not only my voice, but also who I am. I would also like to recognize Lillabeth Broderson and Elizabeth Aaron for their behind-the-scenes work at Ultra Literary.

Thank you to the HarperCollins/Harlequin team for building this book and making it a reality. Their talented editors, John Glynn and Eden Railsback, polished the drafts into final form, while the production team, Gina Macdonald, Katie-Lynn Golakovich, Kristen Salciccia, Amanda Roberts, and Kelsey Heiss-O'Brien, put everything together. Thanks also to VP of publicity, Heather Connor, and senior marketing manager, Abigail Sokolsky, for their efforts.

Finally, I would like to thank my friends and family for their support, especially Jami. As with pretty much everything in my life, I couldn't have done it without you.

Leroy Chiao
November 2025